beschouwde.
een veel an,
offelt waren
litten, t en
jn welk stil
haar vaer,
k de bewe,
r uyt eenige
het bloet
Chyl inde
t men deselve
ende beken.
en mij dit,
volkome

nge, dierk
gehouden,
t sijnder al
t dat wij

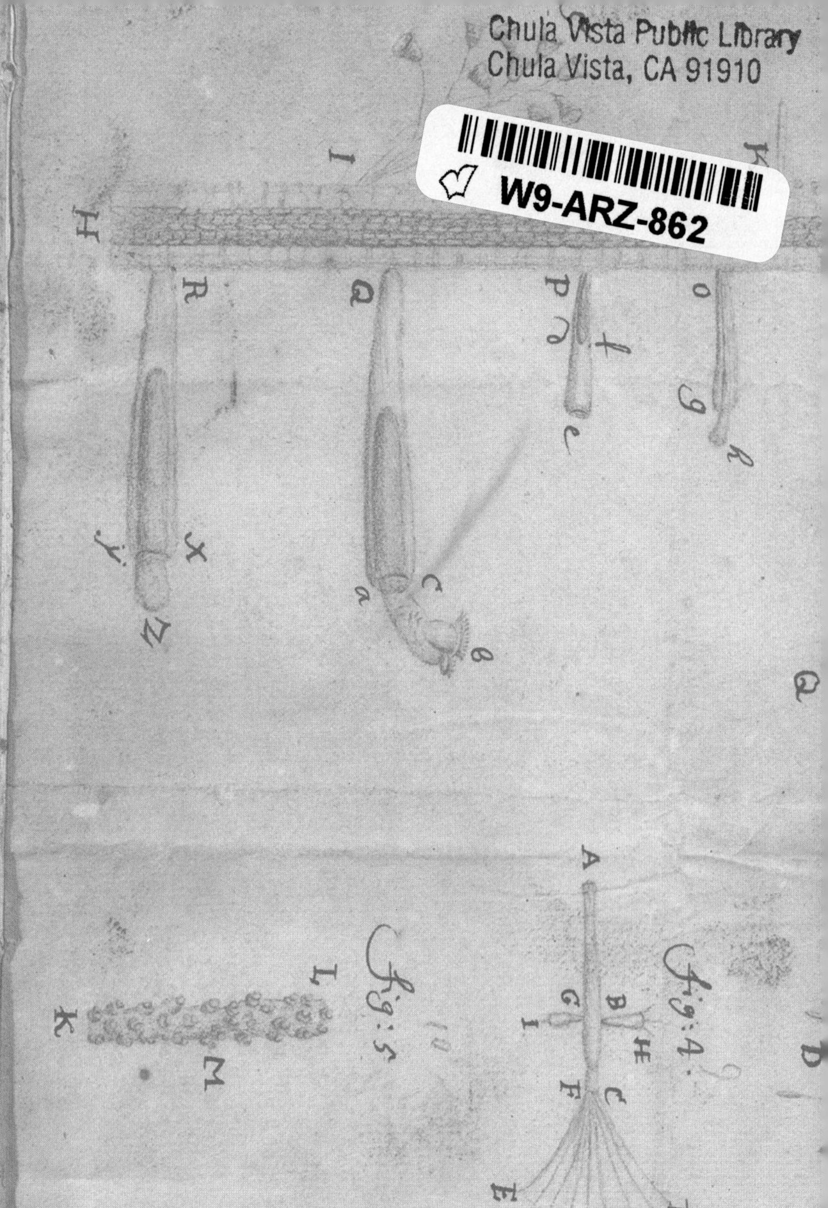

INGENIOUS
PURSUITS

INGENIOUS PURSUITS

Building the Scientific Revolution

LISA JARDINE

NAN A. TALESE

DOUBLEDAY

New York   London   Toronto   Sydney   Auckland

PUBLISHED BY NAN A. TALESE
an imprint of Doubleday
a division of Random House, Inc.
1540 Broadway, New York, New York 10036

DOUBLEDAY is a trademark of Doubleday, a division of Random House, Inc.

Published simultaneously in the United Kingdom by Little, Brown and Company.

Library of Congress Cataloging-in-Publication Data

Jardine, Lisa.
Ingenious pursuits: building the scientific revolution / Lisa Jardine.
p.   cm.
Includes bibliographical references and index.
I. Science, Renaissance.   I. Title.
Q125.2.J365   1999
509.4′09′03—dc21   99-41985
CIP

ISBN 0-385-49325-8

1 3 5 7 9 10 8 6 4 2

For my father, Jacob Bronowski,
who showed me the way,
and for Freya and Zoë, who are the future

CONTENTS

LIST OF ILLUSTRATIONS

CHAPTER 4:   RUNNING LIKE CLOCKWORK

CHAPTER 5:   BREAKING NEW GROUND

CHAPTER 6:   STRANGE SPECIMENS

CHAPTER 8:   COMMITTED TO PAPER

PLATE SECTION 3

ACKNOWLEDGEMENTS

THIS IS A book about the myriad collaborations and exuberant intellectual exchanges that provide the foundation for each and every advance in knowledge. My own researches conform closely to that model, so for a book on the present scale my intellectual debts are inevitably many. Here I select just a few for particular acknowledgement. To all those others, unmentioned, who have helped me in so many ways, my most sincere thanks.

The research on which *Ingenious Pursuits* is based was made possible by my appointment to a visiting interdisciplinary John Hinkley Professorship at the Johns Hopkins University, Baltimore, USA. My thanks to all who encouraged me there, particularly Roger Chartier, Kirstie McClure, Anthony Pagden and Larry Principe. Thanks too to Queen Mary and Westfield College, University of London, for releasing me to take up that appointment, as well as for the support and encouragement of so many of my colleagues there, including most recently our new Principal Adrian Smith. Moti Feingold read the manuscript and corrected some of the slips of the pen.

My editors, Alan Samson in London and Nan Talese in New York, steered the project firmly, yet apparently effortlessly, to completion. Maggie Pearlstine and Toby Green's tireless enthusiasm kept me going whenever my own energy temporarily flagged. Jerry Brotton performed miracles with the picture research — I asked the impossible, and he calmly supplied each elusive image as if the request were simplicity itself.

Of the many librarians and archivists who have given me invaluable assistance, two deserve special mention: Sandra Cummings at the Royal Society in London and Roy Vickery at the Natural History Museum added their own enthusiasm to Jerry Brotton's and mine, and produced even more appropriate texts and illustrations than the ones we were asking for.

My family — John, Daniel, Ali, Rachel and Sam — provided me, as always, with encouragement, unstinting support and a reassuringly calm haven at home. They know how much they contribute to everything I think and write.

London, May 1999

PREFATORY NOTE

A note on dates

TWO CALENDARS WERE in use throughout the period of this book. The Julian calendar was followed in England, and the revised Gregorian calendar was followed everywhere else in Western Europe. The difference between them was ten days in the seventeenth century and eleven days in the eighteenth century (because England observed the year 1700 as a leap year, but the Continent did not). Thus, 12 April in the Julian calendar would be 22 April in the Gregorian before 1700 and 23 April after 1700. In general I have given dates in the form appropriate to the location (Julian in England, Gregorian elsewhere). Sometimes, where correspondence I follow crosses national boundaries (as in the dispute between Huygens and Hooke), the difference in dating becomes significant. In such cases I have specified in brackets following a date whether it is in 'old style' (Julian) or 'new style' (Gregorian).

During the same period the civil year in England began on 25 March. For ordinary usage, however, the new year started on 1 January, as now. Thus, the English civil date 14 February 1675 is 14 February 1676 according to our modern system of dating (some people in the seventeenth century wrote such a date as 14 February 1675/6 for clarity). I have given all dates as though the new year began on I January.

The stories I tell here involve a large cast. The reader will find a brief biographical sketch for each significant character at the end of the book.

INTRODUCTION

IN 1997, SCIENTISTS working at the Roslin Institute in Edinburgh announced a remarkable breakthrough in biological research, in the journal *Nature*. Ian Wilmot and his team had successfully cloned a living sheep using genetic material from cells in an adult sheep's udder.[1] The impetus for this piece of research had come from the rapidly developing field of biotechnology: genetic engineering had already been used to breed sheep whose milk contained vital human proteins used in the medical treatment of cystic fibrosis. Cloning would allow the commercial company, with some of whose funding the research was associated, to produce entire flocks of such sheep, facilitating the production of these new-type medical materials. Ultimately, according to the company's spokesman, Wilmot's cloning technique might be used to 'farm' the human blood clotting factors needed to treat haemophiliacs.

But Dolly the cloned sheep was not heralded as a glorious piece of innovative science. Aghast, the newspapers of the world responded to this sensational scientific advance with a clamour of moral outrage. Driven blindly by the search for the new, we were told, the Scottish scientists were careering towards disaster along that sinister path to damnation notoriously embarked upon by the

I. Dolly the cloned sheep

demonic hero of Mary Shelley's famous novel, Dr Frankenstein. In no time at all we would face the nightmare scenario of genetically engineered armies of identical soldiers, bred to exterminate with ruthless efficiency. Parents would shortly decide exactly what mental and physical characteristics they wanted for their offspring and order them tailor-made, off the shelf. Worst of all, with no further need for sperm in order to beget children, men would be sidelined or cut out of the reproductive cycle altogether, consigned to the scrap heap of history.

For many ordinary people nowadays, it seems, the scientist is the enemy: a detached, remote, forbidding figure, bent uncompromisingly on seeking solutions to complex general problems, without regard for the damaging implications of his 'tampering with nature' for moral probity and human values. The legacy of this inhumane, alien scientific practice is, it is claimed, all around us, beyond the possibility of ordinary people's control. Its malign influence is perceived to lie behind the atom bombs dropped on Hiroshima and Nagasaki, the proliferation of biological weapons in the Middle East, and the experiments routinely performed on animals by cosmetics and other consumer-product manufacturers.

Art and literature, according to this view, stand in humane opposition to science. Artists are the trusted guardians of morality: a small, committed band of civilised and sensitive dreamers who nurture our society's conscience and sustain its values and beliefs. However, in the same year in which Dolly the sheep was cloned, the artist Damien Hirst's *Away from the Flock* — a whole sheep preserved in formaldehyde — was included in the *Sensation* art exhibition at the

2. Damien Hirst's *Away from the Flock*

Royal Academy in London. That exhibition, too, caused outrage with some sectors of the public and censorious remarks from the tabloid press because artists like Hirst refused to give the public pat answers to moral questions. What the critics found disturbing about the exhibition was the perceived cynicism of the art objects displayed, the way Hirst's pickled sheep and the adjacent sliced cow apparently set out with the sole object of shocking their audience. Each unsettled onlooker had to make up their own mind and contribute their own evaluation to the experience of visiting the exhibition.

In the case of art — even avant-garde art — the public and the press expected the artists to be deeply engaged with society and its burning issues, and to offer us telling evaluative insights. Hirst's pickled sheep and segmented cows seemed to refer to animal experimentation and genetic engineering, but refused to offer a moral position. Yet still the very act of provoking us was assumed to make some contribution to a shared understanding of humanity at the end of the twentieth century. Whether they were for or against *Sensation*, those who wrote or talked about it assumed that those who made art were deeply immersed in the world we inhabit, that the ideas and processes of art were tightly interwoven with the fabric of our society's informing ideas and practices. When the hoo-ha was over, public discussion of the works of art in the *Sensation* exhibition turned out to have produced insights into the state of our contemporary society as profound as any stimulated by the more familiarly 'aesthetic' work of early twentieth-century artists such as Cézanne or Renoir.[2]

Personally, in my own intellectual pursuits, I have never felt the need to choose between the arts and sciences. I grew up in a harmonious household in which these 'two cultures' coexisted peacefully. My mother's hands shaped figures out of clay, my father's hands described for us the primitive movements of flint on stone by which 'man the tool-maker' struck fire. At mealtimes, Newton's theory of gravitational pull and the poetry of William Blake were discussed in the same breath; Einstein's and Picasso's enduring contributions to our cultural landscapes were treated as part of a single ferment of intellectual creativity. In that family environment I gained the conviction that imaginative problem-solving is at the root of all human inventiveness, both in the sciences and the humanities. I also learned to believe that intellectually and aesthetically creative people – scientists and artists – were together part of that global community which provides a constantly updated version of those 'human values' on which the well-being of humankind depends.[3]

The scientist is not a malevolent Dr Frankenstein, creating monsters beyond his control. The scientist, like the artist, is one of us. He or she pursues scientific research along directions set by the interests and preoccupations of the community he or she belongs to. What keeps the scientist alert to the moral implications of his or her investigations is that sense of belonging, together with the fundamentally collaborative nature of the scientific project itself. Anyone who has watched a team of scientists at work in a modern laboratory will know that there is more to scientific inquiry than the lonely, rational pursuit of truth. From the designing of experiments to the writing up of results, science is conducted by vigorous group discussion and debate, marked by those moments of dazzling illumination and shared recognition that characterise insight in all domains of human endeavour. Advance in any field has always been preceded by a sudden leap of the imagination, which is recognised for its brilliance by the participating group, and galvanises them in their turn into further activity. Here is a kind of intellectual anthropology that can be further explored. Our Western intellectual heritage has been shaped by ingenuity, quick-wittedness, lateral thinking and inspired guesswork, but not haphazardly. In its detail it is guided and given its informing values by a common code of practice, which is simply an extension of the rules that govern our everyday life.[4]

As ever, those who insist on the separate spheres of art and science claim that the division of arts and sciences is a recent one. Leonardo da Vinci could master the fundamental tenets of ballistics and design a siege-engine for his noble employer as readily as he could handle the colours on his palette to create the enduringly beautiful portrait known as the *Mona Lisa*. Today no single mind has the range and flexibility to cope with both the cerebral rigours of 'hard' science and the mental agility of the contemporary creative arts. This is, I think, to miss the point. I do not believe that science and art are, or ever have been, two distinct practices; rather, they comprise a range of perennially familiar practices in two largely distinct, but occasionally overlapping spheres.[5]

The meeting point – the domain of overlap between styles of ingenuity – is technological inventiveness. It is no accident that the words 'ingenious' and 'engineer' derive from the same root (*ingenium*: mental ability, cleverness, a naturally clever temperament). Within fifty years of Galileo's construction of a working telescope, expert microscopists like Robert Hooke and Antoni van Leeuwenhoek had discovered protozoa and human sperm, using the same lens technology, and were developing the biological theories to explain what they saw. Using the same technology, Dutch artists were producing still-life paintings of unparalleled and lasting beauty. Technology continues today to provide the vital trigger for pure science – it was Maurice Wilkins' X-ray diffraction photographs that set Crick and Watson off on the right path towards discovering the fundamental structure of DNA, and Rosalind Franklin's even more technically exceptional photographs that clinched their argument.

These are the issues I explore in the present book. By temperament a historian, who believes that the clearest answers to our present dilemmas are to be found in the past, I look at the emerging process of science at the key moment of European 'progress'. The changes in intellectual outlook of the sixteenth and seventeenth centuries formed the basis for what many consider to be the most important 'event' in Western history – the so-called 'scientific revolution'.[6] Its breakthroughs in thought and its advances in science still stand today at the centre of every area of modern life. In the fifteenth and early sixteenth centuries, international trade and an increasing demand for consumer 'worldly goods' on the part of the wealthy triggered the European Renaissance in art and learning.[7] The intellectual advances of the scientific revolution took place in the context of the broadened horizons of that consumer revolution. Emerging seventeenth-century science matched and furthered the globalising interests that the Renaissance had stimulated.

The early modern world was in a kaleidoscope of flux before the philosopher John Locke redrew the contours of graspable ideas, and the scientist Robert Boyle heated mercury in his crucible, or Galileo Galilei rolled a ball down an inclined plane. In *Ingenious Pursuits* I explore the forces for change that brought the human and natural sciences together and gave them shape. Each of my selected contributing factors – among them, precise time measurement, enhanced astronomical observation, selective animal and plant breeding, technological advances in navigation, chemical substance analysis, the mathematics of naturally occurring curves – lays a crucial part of the foundations for modern thought and the practice it animates. With each successive layer it becomes easier for the inquiring minds of the next generation to identify the problems on which to focus their attention. The consequences of these key changes surround and envelop them, they shape the world the aspiring mind actively inhabits.

These defining scientific moments were inseparable from the rest of early modern day-to-day life. The key questions whose solution shaped our modern world view arose out of events regarded as remarkable well beyond the confines of any small, select band of intellectual innovators: the appearance of two

unusually bright comets in quick succession; a piece of new technology (such as the microscope) becoming available on the open
market; a new commodity (such as cocoa or coffee) 'taking off'
with discerning consumers; the observed therapeutic effectiveness
of a non-indigenous, naturally occurring substance (such as
rhubarb or quinine).

It is because science grows out of the preoccupations and pressures of everyday life that its discoveries have, in the end, to be
accessible to all of us. Scientific progress ought to be meaningful to
the ordinary person in the street because each of us has participated
in the way of life that has produced the problems pressing for solution. We may not understand the jargon, or know enough
mathematics to follow the equations, but we ought all to be able to
understand that the laboratory that has cloned Dolly the sheep has
thereby made it possible to mass-produce medical products vital to
combat the world's new, virulent diseases.

I begin here with events surrounding the appearance of two
comets in 1680–1. In the course of the story of the 'twinned'
comets, we meet a whole collection of largely unfamiliar intellectual heroes of the scientific revolution – Robert Hooke, Edmond
Halley, John Flamsteed, Jonas Moore, Johann Hevelius, Gian
Domenico Cassini – as well as the better-known figures Isaac
Newton and Christopher Wren. These smaller, 'local' figures are
my 'ingenious pursuers' of innovative understanding. Their
crowded, motley lives, in which conversation in the coffee-house
and vigorous correspondence with like-minded individuals in other
countries figured as importantly as strenuous private study and
laboratory experiment, serve as my model for successful intellectual
engagement in the period. These men crop up repeatedly throughout my story, in the most unexpected locations in Europe, the
East Indies and the Americas, pursuing a curiously varied collection
of investigative goals, and motivated by a volatile mixture of self-
interest, opportunism, curiosity and pure research.

We begin with a collection of technical instruments that became catalysts for scientific advance in the course of the seventeenth century: the microscope, the telescope, the pendulum clock, the balance-spring watch and the air-pump. With the aid of these instruments my ingenious protagonists embarked on strange new quests for understanding the natural world, from the minute detail of the digestive systems of fresh-water shrimps to the regular motions of the farthest visible satellites circling the planets of the solar system. It was in the course of investigations such as these that some justly famous scientific problems – determining longitude at sea, and the functioning of the human vascular and respiratory systems among them – were posed and subsequently solved. Both questions and answers were to large extent instrument- and equipment-driven.

From there we move to mapping, and the scientific disciplines transformed by the extraordinary natural historical discoveries – new plants, animals and minerals – made in North and South America, Africa, India, China and Japan. Here, inevitably, our by now familiar cast of characters operate in an atmosphere tinged with political and military influences, as they pursue astronomical, cartographic, botanical and mineral interests, during unstable political times, in barely colonised territories overseas. Back at home, commercial interests and consumer demands turn out to have had a vital part to play in shaping emerging specialisms as apparently 'pure' as botany and entomology. Meanwhile, such individuals as Hans Sloane and James Petiver, whose considerable personal fortunes were made speculating in new pharmaceuticals like cocoa and quinine, became enthusiastic collectors of rare scientific specimens, channelling their wealth into funding specimen-gathering expeditions, and purchasing the intact collections of other enthusiasts. Men like these ultimately became the founding fathers of our public museums, their largely haphazard assemblages shaping such venerable institutions as London's British Museum.

In a number of ways the process of transmission of the new science was part and parcel of a general explosion in written and printed communication of the same period. That process influenced the very nature of science, and determined some of the future directions of its development. A central part of the activities of the London Royal Society, for example, was the exchanges of letters between the Society's official Secretary and scientific enthusiasts all over Europe. This was the age in which ordinary people assembled sizeable personal libraries; it was also during this period that a number of these were lost in the Great Fire of London.

Finally, I return to the very heart of my argument. The colourful tale of Francis Crick and James Watson's discovery of the structure of DNA – the deep structure of genetics – was told by Watson himself in 1968, in his best-selling book, *The Double Helix*. It was this breakthrough in March 1953 that opened the way to the fundamental advances in understanding of inherited human characteristics currently informing the human genome project, and that led to the genetic engineering which produced Dolly the sheep forty-five years later. That story, I suggest, is uncannily like those that unfolded in the seventeenth century. For the practice of science has been one and the same throughout its history – a story of chance, creative misunderstanding, wrong turnings, sudden opportunities taken, succumbing to sponsorship, and the inspired ingenuity of individual men and women.

CHAPTER I

SIGNS OF THE TIMES

Seeing double

AT THE END of the seventeenth century, a century and a half before the glare of electric street-lighting, the skies above London were dark at night. In that inky blackness, a blazing comet, spotted just before sunrise, early in November 1680, outshining the planets and the familiar constellations of the fixed stars, caused a sensation. A comet of such dazzling splendour would, it was widely believed, bring political and social upheaval in its wake; for a month Londoners observed it through telescopes, tracked its progress across the heavens, and discussed its likely significance over the new beverages (coffee, tea and chocolate) in the coffee-houses. In early December, two weeks after the comet had eventually faded from view, the furore surrounding its appearance intensified when a second, equally bright comet was sighted at dusk, travelling across the heavens in the opposite direction to the first.

Throughout history comets have caused public alarm. For those (and there were many) who regarded comets as ominous portents in 1680, there was much in the way of impending calamities to fix upon. With Oliver Cromwell and the execution of the English King still vivid memories, the stability of the English throne once again looked threatened, this time by the committed Catholicism of Charles II's brother and heir, James, Duke of York (later James II). There had, indeed, been an unusual amount of visible comet activity in the 1660s, suggesting some kind of correlation between the movements of heavenly bodies and England's politically unsettled times. In contemporary doom-mongering broadsides, illustrative woodcuts show one or more comets hanging over the fearful prognostications, charged with responsibility for bloodshed and revolution, the deaths of princes and politicians, drought, famine and flood.

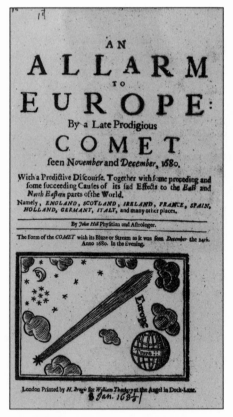

1. Apocalyptic pamphlet about the 1680 comet

2. Five comets seen in England, 1664–82

Two comets in such quick succession caused as much astonishment among London's professional star-gazers as they did with the
general public. From the Greenwich Observatory outside London
the first Astronomer Royal, John Flamsteed, tracked first one and
then the other attentively with the help of his custom-built astronomical telescope, meticulously recording their positions as they
made their way slowly across the heavens. In late December the
second comet was immense, with a tail four times as broad as the
moon and more than seventy degrees long. Flamsteed's colleague,
Edmond Halley, was away from England, crossing the Channel on
his way to visit the French Royal Astronomer, when he glimpsed it
for the first time, 'the tayle of the Comet riseing as it were perpen-
dicularly from ye horizon'.[1] 'I beleive scarce a larger hath ever been
seen,' Flamsteed wrote excitedly to a friend in Cambridge.[2]

Flamsteed was King Charles II's very own 'astronomical observa-
tor' – he had been created the first Astronomer Royal at the age of
twenty-eight, when the Observatory was set up by royal charter in
1675. That founding charter was highly specific as to the purpose
of the Astronomer Royal's appointment. He was to apply 'the most
exact Care and Diligence to rectifying the Tables of the Motions of
the Heavens, and the Places of the fixed Stars, so as to find out the
so-much desired Longitude at Sea, for perfecting the art of
Navigation'.[3] His astronomical expertise, in other words, was to
crack the long-standing problem of determining the precise location
of a sailing vessel on the open sea (a problem that in spite of a large
cash incentive was not solved for another fifty years), and his
appointment followed a Royal Commission set up to give England
the commercial edge in competitive sea-voyaging.

The story goes that Charles II learned from his current mistress
that a Frenchman had devised a method of using the known positions of the moon to determine longitude accurately at sea. When
pressed to produce his lunar method this gentleman proved less
than convincing; nevertheless, the King's attention was drawn to

the fact that accurate star charts were the prerequisite for all such methods of determining position out of sight of land. If England was to seize the maritime initiative and regain international confidence in her supremacy in the trading world, the place to begin, Charles's advisers told him, was to invest magnificently in the latest technology for mapping the heavens.[4]

That the official story (put on the record by Flamsteed himself in his memoirs) makes the establishing of the Greenwich Observatory a consequence of 'pillow talk' betrays the fact that star charts were seen as part of military intelligence, and the new Observatory as a foundation in the national defence interest. Fixing the positions of ships on the high seas with precision was of particular importance for the wartime manoeuvres of the English fleet. Intelligence information (however gleaned) that the French were about to solve the problem was politically alarming. These were years during which England was at war with the Dutch, and when Catholic interests in the form of alliance with the French posed a direct threat to the staunchly Protestant English government. In 1667 the Dutch sailed up the Medway, sank five battleships, and towed the *Royal Charles* back to the Netherlands; the King and his Privy Council were derided for their naval incompetence. Rumours of secret pacts between the Catholic faction in France and Charles II were rife.

National pride was at stake. The French already had their own Observatoire, directed by Gian Domenico Cassini, an Italian by birth, lured to Paris from the University of Bologna in 1669 by the promise of a magnificent annual pension of 9000 livres and a fully funded observatory in which to carry out his astronomical observations. Cassini had moved into the barely completed Observatoire on the southern edge of Paris in 1671 and immediately begun work on an explicitly military-related project, under instructions from Louis XIV's minister for war, Colbert. The Observatoire's astronomers were to survey the whole of France, to produce an accurate map of the royal territories. Straightforward political competitiveness suggested that Charles II too should have an observatory, and should likewise set it to work on some project of national significance.[5]

The leading sponsor of the new London Royal Observatory, and Flamsteed's personal patron, was Sir Jonas Moore, Surveyor-General at the Ordnance Office. The Ordnance Office, located in the Tower of London (which it shared as premises with the Royal Mint), was the seventeenth-century equivalent of the Pentagon, responsible for national security, armaments, military training, and strategic planning in time of war.[6] Moore spotted Flamsteed's potential as a government astronomer (reliable, painstaking and accurate) early in the 1670s; by 1674 he was sponsoring his work, and had promised to pay for the most up-to-date astronomical instruments. Moore personally brokered the establishment of the Greenwich Observatory, and eased Flamsteed into position as its first Astronomer Royal.[7] The Ordnance Office remained responsible for paying the Astronomer Royal's salary down to the nineteenth century.[8]

In June 1675, a group of astronomical enthusiasts led by Robert Hooke and Christopher Wren approved a piece of high ground close to the City as a suitable site for the Royal Observatory. Wren and Hooke, who were both part of the team of surveyors appointed to take charge of the rebuilding of London after the Great Fire of 1666 (nominated by the King and the City of London respectively), designed the Observatory and supervised its building. The nineteen-year-old Edmond Halley, who had been recommended to Moore for his prowess as a skilled astronomical observer, participated in the project from the earliest planning stages (during the same period he joined Moore, Flamsteed and Hooke in observing an eclipse of the moon from Moore's rooms in the Tower).[9]

Sir Jonas Moore took advantage of his senior Ordnance position to ensure that the project was executed with unusual speed and efficiency. Acquisition of the Greenwich site itself was unproblematic, since it consisted of a package of land lopped off the Ordnance's gunnery practice ranges. Construction was put in the hands of the Master General of the Ordnance, who supplied the materials: the building was constructed out of surplus bricks from Tilbury Fort, and iron and lead from a gatehouse demolished at the Tower of London (as part of a general refurbishment of the Ordnance Office there). It was paid for with money raised from the sale of army-surplus gunpowder at Portsmouth.[10]

Once installed, Flamsteed had an impressively large telescope built for himself (paid for by Sir Jonas Moore), complete with cross-wires and a micrometer gauge for added precision. He duly set about compiling a record of astronomical data to outdo those of continental competitors (particularly the French). It was Flamsteed who designated the brick wall on which he mounted a mural quadrant alongside his telescope at Greenwich as the first British zero meridian – the fixed origin for mapping purposes, from which all astronomical and navigational measurements would

FACIES SPECULÆ SEPTEN:

3. Early engraving of the Royal Observatory, Greenwich, showing long telescopes

in future officially be taken. Later he maintained that the ten-foot
quadrant, designed by Robert Hooke, with precision parts by the
master clockmaker Thomas Tompion, was too unwieldy for practical use, but that may have been mere petulance; by then he had
fallen out with Hooke.[11]

When Edmond Halley was appointed the second Astronomer
Royal in 1720, he installed his own, much more powerful, five-foot
transit telescope, creating a purpose-built room for it adjacent the
Great Room in which Flamsteed had worked, and mounting a state-
of-the-art eight-foot mural quadrant (made by clock-maker George
Graham) on a new stone meridian wall. This now became his zero
meridian – 73 inches to the east of Flamsteed's.[12] In 1750 James
Bradley, also Astronomer Royal, picked a third meridian, 436 inches
further east still, in order to accommodate his yet more powerful
telescope. This meridian was adopted by the Ordnance Survey (the
surveying arm of the Ordnance Office), when it started making terrestrial maps in the eighteenth century, and is still taken as zero
meridian for cartographic purposes.

Flamsteed turned out to be as obsessive a data collector as Moore
could possibly have hoped for – a veritable hoarder of measurements. He dedicated his entire career as Royal Astronomer to
recording, hour by hour, night by night, year in and year out, the
positions of the planets and the fixed stars. For forty years he settled the cross-wires of his telescope on each heavenly body in turn,
took its co-ordinates, and inserted it into the monumental body of
data from which ultimately he proposed to build definitive star
maps. Permanently unsatisfied with the accuracy and quantities of
the data he had collected, he endlessly deferred publishing his charts.
When Moore died in 1679, not a single star chart based on
Flamsteed's measurements had yet been produced. In the end, exasperated fellow-astronomers, desperate for the information
Flamsteed had collected, had clandestinely to wrest the tables from
his grasp. In 1712, the greatest theoretical astronomer of them all,

Isaac Newton, who was by then President of England's most influential scientific institution, the Royal Society (the 'Royal Society of London, for Improving of Natural Knowledge', to give it its full title), issued instructions that Flamsteed's *History of the Heavens* should be published without the consent of its author – indeed, with Flamsteed protesting vigorously that further emendation and correction were still needed before the star maps could properly be printed.

Newton's argument was that Flamsteed's salary was paid from the public purse, and that all his work therefore belonged in the public domain. Flamsteed's riposte was that the equipment which gave access to the contested data was his, built and paid for for his own personal use. (Flamsteed won a technical victory: in 1715 he gained authorisation to recall and destroy the three hundred copies of *History of the Heavens* that had not yet been sold. When Halley arrived as the new Astronomer Royal after Flamsteed's death he found that Flamsteed's widow had removed all the astronomical instruments, on the grounds that these were his predecessor's personal effects.)[13]

Star-gazing at home and abroad

Flamsteed knew the map of the heavens like the back of his hand. In 1680 a shining stellar newcomer was as instantly striking as an extremely rare plant, spotted in an unlikely habitat by an enthusiastic botanist. And, like the botanist, Flamsteed's next instinct after locating his comet was to ask himself whether this was a chance occurrence, or whether he might have anticipated it. In early 1681, Flamsteed decided that what had appeared between early November and late December was not two comets but one – a single comet moving in a straight line, which had abruptly changed path in front of the sun, and was now travelling in a straight line in the opposite direction. In fact, in an uncharacteristic burst of ingenuity he proposed that magnetic forces had at first drawn the comet towards the sun, then repelled it. It was a bold idea. He had his recorded observations; what he needed was an outstanding mathematician to find the means of checking his hypothesis against the data. Through a Cambridge University friend, he approached the brilliant (but as yet relatively unknown) mathematician Isaac Newton at Trinity College.

Newton was not a man to humour fellow scientists. He was notoriously quick to take offence himself, and when he regarded others' proposed explanations of natural phenomena as implausible he didn't hesitate to say so. He had little respect for Flamsteed's scientific ability, regarding the Astronomer Royal (with some justification) as a glorified information-gatherer, not a thinker. But Flamsteed's inquiry rekindled his own long-standing interest in the comets. Newton took some observations himself, as the second comet moved through the heavens nightly in the early months of 1681.[14] He read the most recent continental papers on comets, and collated all the available data of observed comets' paths that he could gather together.[15] And he convinced himself to his own satisfaction that Flamsteed was mistaken: that there had been two comets not one, and that the idea of a comet that stopped dead in its tracks in front of the sun and reversed its direction was untenable. He wrote to Flamsteed in April 1681 sternly telling him so: 'To make the comets of November and December but one is to make that one paradoxical.'[16]

Years before Flamsteed's inquiry, comets had played an important part in Newton's developing fascination with the movements of heavenly bodies. One of the few surviving anecdotes from Newton's childhood describes a practical joke he played as a schoolboy at Grantham Grammar School in the 1650s. 'He first made lanterns of paper crimpled, which he used to go to school by, in winter mornings, with a candle, and tied them to the tails of the kites in a dark night, which at first affrighted the country people exceedingly,

4. Newton in later life

thinking they were comets.'[17] Newton's prank was apparently the talk of Grantham, the locals discussing earnestly the calamitous events the 'comet' would visit upon them.

Emotionally deprived and fatherless, young Newton was (on his own confession) more inclined to 'peevishness with [his] mother, or 'striking [his] sister' than practical joking.[18] The fake comet story suggests that the drama of comets' ominous appearances captured his imagination from an early age. At Cambridge, as he later recalled, he 'sate up so often long in the year 1664 to observe a comet that appeared then, that he found himself much disordered and learned from thence to go to bed betimes'.[19] His memory served him false about the lesson he learned from his first closely observed comet, however. Throughout his life, lack of sleep brought on depression and nervous collapse when he was distracted from the mundanities of everyday living by a burning intellectual issue.

It is remarkable how many Renaissance mathematicians chose to date their interest in science from a key moment involving a comet. In old age the German astronomer Johannes Kepler – the man who clinched mathematically the fact that Copernicus had been right in believing that the earth revolved around the sun – singled out from his memories of an otherwise appallingly deprived childhood the day when he was 'taken to a high place' by his mother to look at the comet of 1577 (he was six years old).[20]

Sir Christopher Wren was another who kept a nightly vigil, this time fascinated by the comet of 1664. An able mathematician, Wren presented the Royal Society with a geometrical method for calculating the true path of the comet, which he judged to be a straight line away from the sun, its movement retarded by a magnetic attraction towards the sun, which eventually causes it to fall back towards it.[21]

Having dismissed Flamsteed's idea as implausible, and abruptly terminated their correspondence, Newton nevertheless continued to mull over the curious phenomenon of the two consecutive comets. He was further intrigued by the appearance in 1682 of yet a third – the comet known today as Halley's. He collected together what data for this comet he could lay his hands on; he tracked the 1682 comet himself. And some time between 1682 and 1684 he changed his mind: the 'two' comets of 1680–81 had indeed been one – a single 'sun-hugger', turning tightly in a parabolic orbit around the sun, before setting off again into distant space.

Newton never had the good grace to apologise to Flamsteed for having initially poured scorn on the Astronomer Royal's idea that the comets of November 1680 and December 1681 were one and the same. Long afterwards, when Newton was taking the credit for establishing the sharply elongated elliptical orbit of the 1680–81 comet, Flamsteed continued to grumble that 'this is what he before contended against with some virulency, but he [Newton] had no mind to remember it'.[22] Flamsteed's later refusal to release his star charts into Newton's hands for publication was no doubt coloured by his earlier run-in with him.

Here the matter might have rested, as what Flamsteed termed an 'amicable controversy' between fellow-astronomers, had it not been for the intervention of a young man of a much bolder temperament. Edmond Halley was a ground-breaker, someone who liked to use his ingenuity to devise his own projects and conduct all aspects of their exploration himself. He was as gregarious as Flamsteed and Newton were solitary; as widely travelled as they were reclusive; as easy-going as they were both difficult. He was a hands-on practitioner of a whole range of activities connected with long-distance sea-travel, map-making, commerce and cargo-salvaging. The annals of the Royal Society are littered with enterprising papers by Halley on everything from the global patterns of trade winds, to the mechanics of diving bells, the rise and fall of mercury in the barometer, compass variation, and the beneficial effects of opium-taking. Not for Halley the detached life of the university don or the monotonous routine of the observatory. In the 1680s the studious Flamsteed complained loudly that Halley drank brandy, smoked, and swore like a sea-captain.[23]

At the outset of his career, Halley had talked his way into a job working with Flamsteed on his comprehensive star-measuring at Greenwich, by writing to him to point out glaring errors in the current star charts. It was soon clear, however, that at Greenwich he would be permanently overshadowed by more assiduous data-collectors. So in 1676, having spotted an opportunity to make technological advances that would make more of an impact, he again took the initiative, and approached the Royal Society for backing. 'I would willingly do something to serve my generation, and here I can do nothing but what will be rendered wholly inconsiderable by the greater accurateness of the great promoters of astronomical science of our age,' he wrote.[24] His alternative proposal was that he be sent to map the stars in the southern hemisphere for the first time since precision measurement had become possible – stars for the most part not visible from Greenwich's latitude, and never before observed with the hi-tech help of a telescope.

Halley's project had strategic implications, which were immediately evident to the entrepreneurially minded in London's rapidly expanding commercial community. Fixing the locations of stars in the southern hemisphere was crucial for sea-traffic to and from the lucrative markets in the East, as ships tried to find their bearings as they navigated down the west coast of Africa, out of sight of the pole star, and towards the treacherous Cape. Charles II took a personal interest, and recommended Halley's enterprise to London's foremost trading conglomerate, the East India Company. They promptly offered Halley facilities on the island of St Helena, the Company's permanent base in the southern Atlantic Ocean (granted to them under charter by Charles in 1661 and again in 1673) from which it supplied its ships on their return voyages from the East Indies, and the only English possession south of the equator. The Company also offered Halley free passage there and back with his equipment and an assistant.

Between February 1677 and March 1678 Halley determined the positions of 341 stars, and sent home a steady stream of scientific reports.[25] Returning to England (after three months at sea on the East India Company's ship, the *Golden Fleece*), he published a

EDMVND. HALLEIVS LL.D.
GEOM.PROF. SAVIL. & R.S.SECRET.

5. Edmond Halley

catalogue of telescopically determined star positions, and a large celestial planisphere comprising star maps for both northern and southern hemispheres. Actually, the adventurous nature of the project was more impressive than its execution (Flamsteed was critical about the comparatively small number of observations on which

JOHANNIS HEVELII
COMETOGRAPHIA.

Andr. Stech delin.

E. Nscher Sculps.

6. Aristotle, Hevelius and Kepler dispute the orbits of comets

Halley had based his planisphere, and sceptical about their accuracy). Persistent low cloud over St Helena – a climatic feature upon which visitors regularly commented – hampered Halley's observations; we might suspect that he was in any case rather less single-minded than Flamsteed to spending his entire waking life training his astronomical instruments on the stars. (Temperamentally, Halley preferred intensive observation over short periods, to verify a hypothesis he had already arrived at, rather than steady observation over time.)

Halley made good use of his time on St Helena, even when the skies were overcast. He stayed in touch with his Royal Society sponsors throughout his year-long absence, reporting back via Sir Jonas Moore on the scientific experiments he carried out there: using a mercury barometer he tested Wren's theory that 'the pressure of the air was not the same in all parts of the earth, but in some places always more in some always less';[26] and he recorded magnetic variation (the deviation of a compass needle from true North). During a partial break in the clouds he observed a transit of Mercury (the passage of the planet across the disc of the sun, observable only rarely), sending home his measurements to be used alongside those taken of the same event by French astronomers at Avignon.[27]

Expert opinions

In 1678, Halley returned in some triumph from St Helena and rushed his new star charts into print. Publication was a thoroughly in-house affair. His assistant on the St Helena voyage, James Clerk, engraved the plates, and his friend Hooke (who had experience with map-printing) also helped with production.[28] Halley dedicated his southern star map to the restored monarch, King Charles II, and named one of his new constellations 'Charles's Oak' after the oak tree in which Charles had hidden at the battle of

Northampton.[29] For the matching star map of the northern hemi-sphere he used the great astronomer Tycho Brahe's star positions (since Flamsteed's new measurements were not yet available); he shrewdly dedicated this to Sir Jonas Moore.

Halley had succeeded in making a name for himself with the com-mercial community and with the Ordnance Office, both of whose interests he served assiduously for the rest of his long life. But his talent for self-promotion and spotting good business opportunities (along with his somewhat cavalier attitude to precision measuring) led to a rapid cooling in relations between himself and Flamsteed. In spite of their close early relationship, by the 1690s, when Halley badly needed access to Flamsteed's scrupulous astronomical data to clinch his theory concerning returning comets, the two men were no longer on speaking terms. Though Flamsteed insisted that what he objected to was Halley's increasing preoccupation with funding, their estrangement was directly related to their scientific sponsorship. Sir Jonas Moore and the Ordnance Office were backers for both Flamsteed's and Halley's star charts, and footed the bill for the expenses associated with them. Flamsteed was unhappy that funds which might have gone towards purchasing further cutting-edge astro-nomical instruments for Greenwich were diverted to pay for Halley's equipment, travel and expenses for the St Helena venture.

Designing and naming new constellations to flatter a patron was a favourite astronomer's ploy, guaranteed to please. Halley's con-temporary, the arch-royalist Edward Sherburne (Clerk of the Ordnance Office), inserted Charles I's bleeding heart as a slightly tasteless new constellation in his English star chart.[30]

A generation earlier when Galileo discovered the moons of Jupiter through his telescope, he named them the 'Medicean stars' in honour of his patron (and tried to recruit the Medici merchant-bankers to market his telescopes and his treatise on astronomical observation).[31] Johann Hevelius of Gdansk placed the shield of his patron Jan Sobieski, King of Poland as a constellation on his star

7. Halley's southern star map showing 'Charles's Oak' constellation

map. He also retained Halley's 'Charles's Oak' in his southern hemisphere chart, as a compliment to Halley.[32] Halley had visited
Hevelius's private observatory at Gdansk in 1679, sent as a semi-
official representative of the Royal Society to observe his method of
angular measurement of the positions of stars with the naked eye

(rather than with the aid of the new telescopic sights). Halley tes-
tified that Hevelius's exceptional eyesight and skill with his sextant
and quadrant gave him a measuring accuracy as good as any gained
using a telescope – though later, in correspondence with Newton in
the 1690s, he pronounced Hevelius's observations and calculations
wholly untrustworthy.[33]

Hevelius's astronomical work was a kind of yardstick by which
other European astronomers measured their own accuracy. In the
1660s he had been singled out by Louis XIV's minister for war,
Colbert, as the leading practitioner in Europe, and allocated a grant
from the French treasury – a gesture Hevelius repaid by dedicating
his first book on comets to Colbert, and two later works to Louis
XIV.[34]

The business side of Halley's life involved him from this point on
in a whole series of undertakings to assist the Ordnance Office in
comprehensively mapping the globe for strategic purposes. The
public stature of 'Captain Halley, FRS, Royal Navy' grew still further in the late 1690s, when he took command of a small ship, the
Paramore, and undertook three ambitious voyages, jointly sponsored
by the Navy and the Royal Society, to chart global magnetic variation. The Admiralty instructions issued with his commission (which
Halley drew up himself) stipulated that he was 'to endeavour to gett
full information of the Nature and Variation of the Compasse over
the whole Earth, as Likewise to experiment what may be expected
from the Severall Methods proposed for discovering the Longitude
at Sea. In all the Course of your Voyage, you must be carefull to
omit no opportunity of Noteing the variation of the Compasse, of
which you are to keep a Register in your Journall'.[35]

Halley made about 150 observations of magnetic variation over
the Atlantic, under extremely difficult circumstances (he records
the hair-raising weather they encountered in his journal, alongside
his measurements). Shortly after his return, in 1700, he published
his measurements in the form of a map – the first map to use

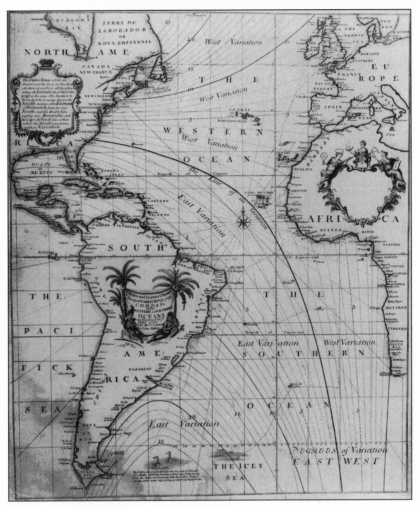

8. Halley's isogonic map of magnetic variation, showing St Helena

isolines (lines connecting points of equal intensity of phenomena whose intensity varies with geographical location).[36] It hung in the Royal Society's meeting rooms, providing impressive evidence of the Society's mastery of matters crucial for navigation, and firmly establishing its author as the foremost figure in scientific surveying.

Two years later, Halley produced an equally original map of the English Channel and its tidal movements (which also allowed some strategically important, covert surveying of the coast of France). The habits of mind developed during Halley's daring youthful journey of 1676 to pursue astronomy on a volcanic rock in the middle of the Atlantic shaped an unusually wide-ranging, global mentality in the mature man.[37]

No one knows what gave Edmond Halley the bright idea in 1684 of going in person to Cambridge to consult Isaac Newton about the trajectories of comets. Halley, too, had been struck by the December 1680 comet's appearance (the 'second' comet of which Flamsteed had exclaimed that 'scarce a larger hath ever been seen'), which he observed further with Cassini at the Paris Observatoire (Cassini gave him a copy of his own treatise on comets as a gift).[38] His friend Robert Hooke had published on comets; his Royal Society colleague Christopher Wren was absorbed with the mathematical calculations needed to prove that the 1680 comet, like its 1664 predecessor, had moved in a straight line. Later Halley told Newton that he, Hooke and Wren had discussed the inverse square law as the defining feature of planetary orbits together in a coffee-house in early 1684 – but that was in order to reassure Newton that the litigious Hooke could not claim personal priority for an idea that had emerged out of the Royal Society's collaborative discussions.

At any rate, it was entirely typical of Halley to decide to cut loose from the parochial, and increasingly unproductive, speculations about comets' paths that were going on in London, and to make the imaginative leap elsewhere – just as he had set off impetuously in pursuit of unmapped constellations six years earlier. The question to which he needed the answer was deceptively simple. If, as he and his fellow-astronomers were coming grudgingly to agree, there was an attractive force between the planets and the sun, inversely proportional to the square of the distance between them, what then was the equation for the planet's motion?

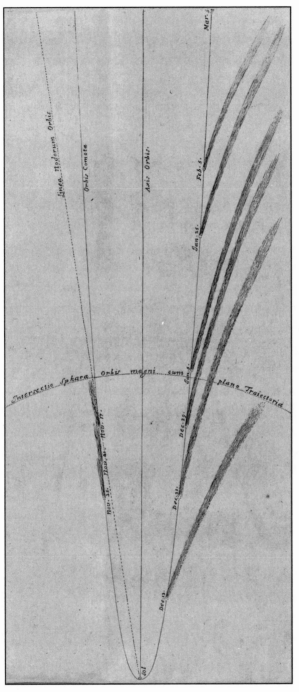

9. Newton's sketch of the orbit of the 1680 comet

Newton, as we know, had gone on thinking about the orbits of heavenly bodies ever since Flamsteed's first inquiry. When Halley asked him if he had any idea in what regular way a planet really moved, Newton replied, with studied nonchalance, that 'naturally' it would be an ellipse – he had once written a paper proving it, though he could not at this moment lay his hands on it to lend to Halley. Halley was beside himself with excitement. Furthermore, Newton told Halley that he could prove that the inverse square law implied an elliptical orbit around the sun as one of its focuses. Comets, he added, in response to Halley's inquiries concerning the double phenomenon of 1680–81, were, to all intents and purposes, 'a kind of planet'.

If Newton had had his way, the conversation with Halley would merely have produced one of those ingenious academic communications, recorded with so many others, on all kinds of contemporary brain-teasers, wild theories and practical inventions, in the Proceedings of the Royal Society. As promised, he sent a brief paper to Halley explaining his grounds for believing that he could calculate an orbit for the 1680–81 comet's path between November and the following March based on Flamsteed's observations (his suggestion at this stage was a parabola). But Halley was not satisfied. He was both persistent and intellectually astute enough himself to recognise the extraordinary significance of Newton's mathematical breakthrough. He insisted that a fuller discussion would be well received by the Royal Society. As its clerk, producing the minutes of meetings, he engineered wholehearted backing on the Society's part for a book-length explication of Newton's mechanics of planetary orbits (including financing the publication).

It took three years for Newton to complete his *Principia* (*The Mathematical Principles of Natural Philosophy*). The theoretical principles revealed in that work were entirely his, the unique product of a brilliant mathematical mind. But the book, and above all the practical use of astronomical data it incorporated, was a collaborative effort,

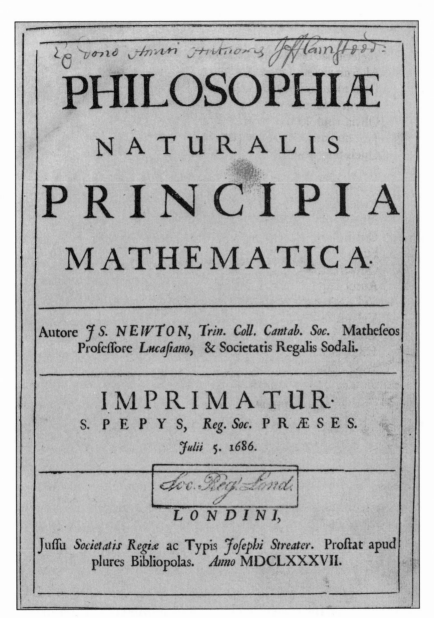

PHILOSOPHIÆ

NATURALIS

PRINCIPIA

MATHEMATICA.

Autore *JS. NEWTON*, *Trin. Coll. Cantab. Soc.* Matheseos
Professore *Lucasiano*, & Societatis Regalis Sodali.

IMPRIMATUR.
S. PEPYS, *Reg. Soc.* PRÆSES.
Julii 5. 1686.

LONDINI,

Jussu *Societatis Regiæ* ac Typis *Josephi Streater.* Prostat apud
plures Bibliopolas. *Anno* MDCLXXXVII.

10. Title page of Newton's *Principia* – Flamsteed's copy

co-ordinated by Halley. Halley extracted the manuscript in sections from its author, wheedling, flattering, reassuring and chastising him, to get him to give up the pages for printing. Hooke's quarrelsome intervention, claiming that the inverse square law belonged to him, prompted Newton to threaten to withhold the entire third book – the crucial part containing the calculation of the 1680 comet's orbit. Halley, through a combination of flattery and fibs, eventually persuaded him to change his mind.

The whole process took so long that the Royal Society lost interest in the project. Its Council had also had their fingers burned in the meantime with an ambitious, lavishly illustrated edition of Francis Willoughby's *History of Fishes* (1686) (a work completed after the author's death by the botanist John Ray), which cost a fortune to produce, and failed to sell. By 1687 Halley had to take the cost of the entire *Principia* publishing venture upon his own shoulders. He was also obliged to accept his entire remuneration as newly appointed Clerk to the Royal Society (a post for which he was in competition with the future founder of the British Museum, Hans Sloane) in the form of copies of the *History of Fishes*, at a nominal value of £1 a piece.[39] Fortunately Halley's various seafaring ventures had by this time made him independently wealthy enough to be able to absorb both financial burdens.

The collaboration between Newton and Halley was a crucial component in the final scientific breakthrough that Newton's *Principia* represents. Taking the measurements to record the successive locations of any moving celestial body required technical skill and the latest equipment. Hypothesising that comets, like the planets, might move along some kind of non-linear path, some kind of conic section, required a keen sense of intellectual possibilities, and a good grasp on the current literature. But proving the matter mathematically required genius of the kind the socially dysfunctional Newton alone possessed. While the comet moves along its celestial trajectory, the earth, too, moves along its own elliptical orbit, in a different geometrical plane. Calculating the former from observations taken on the latter, therefore, required a method of successive approximations (which Newton describes in his *Principia*). Even today it is difficult to see how these approximations could have been made on the basis of a comparatively few observations.

Halley's own most important contribution to the theory of comets came directly out of the many long months he spent cajoling Newton into completing the manuscript of his *Principia*. Following the remarkable success of the first edition, Halley became concerned to regularise the method of successive geometric approximations that Newton had used for computing orbits from observations in Book Three. In 1695 he wrote to Newton offering himself to take on the time-consuming task of working through all the available measurements, recalculating algebraically, and correcting for observational errors, 'to ease you of as much of the drudging part of your work as I can, that you may be better at leisure to prosecute your noble endeavours'.[40] Three weeks later he notified Newton that he had worked through the observations on which Newton had based the orbit of the 1680–81 comet in *Principia*, and 'I find in that of the 25the of January 1681, that there is a mistake of 20 minutes in the Longitude that day, or 56 minutes for 36, and so I have it in a letter Mr Flamsteed sent me when I was in Paris' (in other words, Flamsteed had made an error of transcription in transmitting his data to Halley). Once that correction is made, Halley continues, the conic section that best fits the observations is an ellipse, and 'I am satisfied that it will be very difficult to hitt it exactly by a Parabolick.'[41]

In late October 1696, Halley told Newton, in a further letter, that as a result of his own calculations, using Flamsteed's observations of the 1682 comet (obtained via Newton, since Flamsteed and Halley were not on speaking terms), he was convinced that the comets of 1607 and 1682 were one and the same. An entry in the Journal Books of the Royal Society for 3 June 1696 records that:

Halley produced the Elements of the Calculation of the
Motion of the two Comets that appear'd in the Years 1607
and 1682, which are in all respects alike, as to the place of their
Nodes and Perihelia, their Inclinations to the plain of the
Ecliptick and their distances from the Sun; whence he
concluded it was highly probable not to say demonstrative, that
these were but one and the same Comet, having a Period of
about 75 years; and that it moves in an Elliptick Orb about the
Sun, being when at its greatest distance, about 35 times as far
off as the Sun from the Earth.[42]

Halley had found 'his' comet.

At the French Observatoire, the reception of Newton's *Principia*
was lukewarm, and there was considerable scepticism over the inverse
square law of planetary attraction on which Newton's theory of
cometary orbits also depended. Nicolas Fatio de Duillier, who was
in England in 1687, wrote to Christiaan Huygens:

> I have been three times to the Royal Society where I have heard
> both very good proposals and some platitudes. Some of those
> gentlemen have an extremely favourable prejudice about a book
> from Mr Newton that is now in press and will be issued in
> three weeks. I have seen parts of the treatise, and it is certainly
> full of many valuable propositions, but I wish, Sir, that the
> Author had taken some advice from you about this principle of
> attraction he assumes between celestial bodies.[43]

French astronomers were most interested in the tabulated results,
which had been Halley's responsibility, and which included data
compiled by French astronomers and supplied to Halley by Cassini.
Halley presented one of the very first copies of *Principia* off the
presses to Cassini, with the inscription, 'presented in great humility
as a token of a grateful heart'.[44] From Halley's and Cassini's point
of view, Newton's work was a tribute to an on-going collaboration
between the best astronomical observers from two Royal

Observatories. The theoretical work was secondary, and represented Newton's attempt to 'make sense of' the data. It was the nitty-gritty of practical observation, and the tabulated movements of celestial bodies based upon it, that impressed the international community, and meant that Halley's reputation as an astronomer at the end of the seventeenth century surpassed that of Newton.

The shock of a comet

Great ideas are the product of collisions of minds and broken boundaries. Flamsteed's and Halley's practical astronomical interventions provoked Newton into theoretical consideration of cometary motion, and then to propose a universal attractive force between heavenly bodies, whose consequence was the elliptical orbits of planets and comets alike. But Newton's 'force of gravity' was far from being a piece of pure mathematical rationalising of the data. The idea of this 'occult' invisible attraction owed as much to the cosmic mysteries appealed to by more superstitious men to justify the improbable frequency of appearance of comets during the latter half of the seventeenth century – a period of unusual political instability and unrest across Europe.

In the first edition of *Principia*, Newton hinted that he believed the harmony and balance of the universe might be coming to an end, and that one of the spate of remarkable celestial phenomena recently witnessed might signal the approaching end of the world. In 1697 Halley read a paper to the Royal Society in which he described the effects 'of a Collision of a great Body such as a Comet against the Globe of the Earth', whereby 'that Earth might be reduced again to a Chaos'.[45] For both men, the comet of 1680–81 was particularly significant. When he was in his eighties, Newton told his nephew John Conduitt that he believed the earth had narrowly escaped destruction by the 1680 sun-grazing comet. Had that comet been drawn into the sun, the earth would have been destroyed in the ensuing conflagration. Halley agreed. The 1680 comet, he wrote in the revised version of his *Synopsis on Cometary Astronomy* (1705):

> came so near the path of the Earth, that had it come towards the Sun thirty-one days later than it did, it had scarce left our Globe one Semidiameter of the Sun towards the North: And without doubt by its centripetal force (which with the great Newton I suppose proportional to the bulk or quantity of matter in the Comet), it would have produced some change in the situation of the Earth's Orbit.[46]

Both Halley and Newton believed that eventually, on some future occasion on which the 1680 comet returned (its period of rotation was 575 years, according to Halley's calculations), 'the shock of a comet' would bring the world to an end. Newton explained to Conduitt:

> He could not say when this comet would drop into the sun; it might perhaps have five or six revolutions more first; but whenever it did, it would so much increase the heat of the sun, that this earth would be burnt, and no animals in it could live.[47]

Newton and Halley believed devoutly in a world of awesome complexity, designed and regulated by an all-powerful, all-knowing God. It was consistent with those beliefs to maintain that comets moved in closed orbits, and to predict, as Halley memorably did, that a further comet seen in 1682 would return in 1758. It was equally consistent to believe that the newly discovered periodicity of cometary orbits presaged the ultimate end of the world in an inferno caused by a comet (Newton and Halley were inclined to believe that such a collision had heralded the beginning of the world, also).

In the realm of scientific progress, history yearns for simple stories, told with clarity, of great men spontaneously solving the long-standing puzzles of the natural world. But Newton needed both Flamsteed and Halley to jolt him into the idea of cometary orbits. And the idea could never have taken hold if it had too abruptly overturned the customary world view. To take imaginative hold on its times, a great idea needs a measure of compatibility with the most deeply cherished of contemporary beliefs. The story of the way in which the motion of the 1680–81 comet came to be understood is a story of gradually modifying views of the cosmos, networking and teamwork, sudden insights and occasional errors of judgement, culminating in the intellectual landmark of Newton's *Principia*.

·CHAPTER 2·

S CALE M ODELS

Through a glass darkly

ON A WINTRY day in January 1665, in London, Samuel Pepys, Secretary to the Admiralty and keen amateur scientist, brought home a copy of Robert Hooke's newly published *Micrographia* – the seventeenth-century equivalent to a coffee-table book. The volume featured Hooke's breathtakingly beautiful engravings of a whole range of natural phenomena seen under the microscope. Strange flora blossomed on the page before the enchanted Pepys's eyes; a handful of thyme seeds took on the appearance of a Dutch still-life of lemons. He could not put the book down, noting in his diary that the following evening he 'sat up till 2 a-clock in [his] chamber, reading of Mr. Hooke's Microscopical Observations'. It was, he recorded, 'the most ingenious book' he had ever read.[1]

The pictures of many-times magnified objects in *Micrographia* were more impressive than anything Pepys had seen in the course of his own amateur dabblings in microscopy. Like other gentlemen-about-town he had recently purchased a microscope himself – the latest thing in high-tech playthings. It had cost him five pounds ten

I. Engraving of much-magnified thyme seeds from Hooke's *Micrographia*

shillings from Richard Reeve, the London instrument-maker, and came complete with plane glass slides of prepared specimens.[2] But when Pepys peered down his microscope he was dismayed to find he saw nothing at all (neither could his wife do any better). Eventually, after repeated attempts at adjusting the instrument, they were able to make out dim, magnified images, 'with great pleasure, but with great difficulty before we could come to find the matter of seeing anything by my Microscope – at last did, with good content, though

not as much as I expect when I come to understand it better.'[3]
Looking down a microscope, like looking through a seventeenth-
century telescope, took practice.

Pepys probably took heart from the fact that the author of the
Micrographia himself admitted in his preface, that 'the Glasses [lenses]
of our English make, though very good of the kind, are yet far
short of what might be expected', and that 'after a certain degree of
magnifying, they leave us in the lurch'.[4] Hooke himself, he confided
to the reader, had encountered problems when he settled down
with his 'interpreter's glass' to observe such entertaining micro-
phenomena as head lice, fleas, mould and the minute worms to be
found in seventeenth-century vinegar. In order to make the exquisitely detailed drawings upon which *Micrographia*'s illustrations were
based, it turned out that he needed a brighter light-source than sun-
light or candlelight to illuminate his specimens, to sharpen the
quality of the image seen through his microscope eyepiece.
Undaunted, and with characteristic improvisatory flair, Hooke had
invented a purpose-built piece of image-enhancing apparatus. He
placed a light-condensing brine-filled globe between his lamp light-
source and his specimen, then narrowly focused the lamp's
intensified beams by means of a convex lens. By adjusting the relative positions of lamp, globe and lens he found he could improve
the magnified image of his subject considerably. He called it his
'scotoscope'.[5]

Such a hands-on solution is typical of Robert Hooke. A man of
resolutely practical bent, his response to any scientific problem was
to invent a piece of equipment to resolve it. Hooke was one of
seventeenth-century science's most accomplished experimentalists.
To gain knowledge and mastery of nature (of a kind likely to make
a man's fortune in sponsorship and patents), all that was needed,
in his view, was technical know-how and a devoted band of skilled
craftsmen to execute the instruments. 'My ambition is, that I may
serve to the great Philosophers of this Age, as the makers and

grinders of my Glasses [lenses] did to me,' he declares in the preface
to *Micrographia*. His approach gained endorsement from some formidable contemporary figures: in a polemical treatise on the human
understanding and the sense organs, Thomas Hobbes, best known
for his political treatise *Leviathan*, chose Hooke's microscope light-
condenser as an example of the way the limits of the human senses
could be extended by experimental improvisation.[6]

Hooke's *Micrographia* begins pragmatically with a set of practical
instructions for microscope users. Its first engraving shows a commercially manufactured microscope, readily available on the London
market from the workshop of Christopher Cock, although the text
of Hooke's preface explains pretty clearly that he has had to modify
and improvise with such microscopes to achieve his own remarkable
images. Commenting on the current state of telescope technology,
he recommends Richard Reeve's instruments by name – providing a
puff in print for him, and sending his readers off in that direction
if they hankered after an enlarging instrument of their own. This
coffee-table book was also a user's manual. (Pepys had earlier bought
the physician Henry Power's book on microscopy, and found it
unhelpful – it had no pictures.)

Richard Reeve was indeed one of the foremost instrument sup-
pliers making state-of-the-art lenses and appliances in England and
beyond in the 1650s and 1660s. In 1655 Christopher Wren praised
Reeve 'who makes the best of any microscopes to be had'.[7] Dr
Henry Power owned a Reeve microscope; so did Seth Ward's observatory at Oxford. Reeve was also the foremost English producer of
lenses for telescopes. In 1664 the secretary of the newly established
Royal Society, Henry Oldenburg, wrote with satisfaction (and not
a little propaganda salesmanship for the Society) to the astronomer
Hevelius in Gdansk: 'We have constructed a telescope 60 feet long,
recently finished by our countryman Reeve, whose object glass of
about five inches diameter bears an aperture of 2⅙ inches.'[8] After a
certain amount of haggling over the price of this 'excellent' object

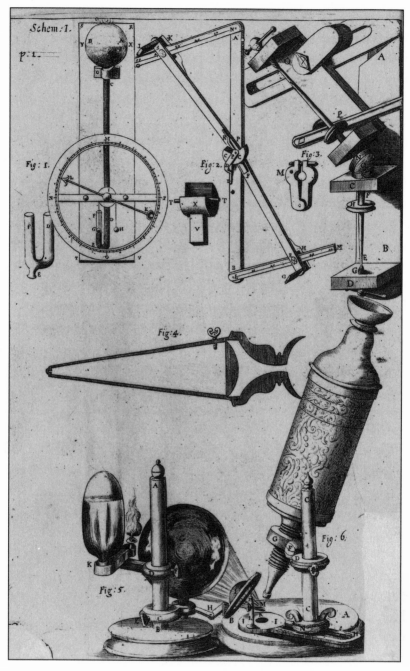

2. Hooke's microscope and 'scotoscope'

glass (at one stage Reeve was valuing it at £100), the cost was borne by Robert Boyle.

The Reeve shop provided sophisticated technical after-sales support to his customers too. In 1661 the celebrated Dutch scientist Christiaan Huygens, who was on his first trip to London, observed a transit of Mercury from Reeve's shop, together with the London astronomer Thomas Streete (the transit occurred, auspiciously, on Charles II's coronation day). In his *Astronomia Carolina* (1664) Streete describes having used 'a good telescope, with red glasses for saving our eyes'. Huygens had sent for a lens from home some time earlier – perhaps Reeve had used it to construct a suitably powerful telescope.[9]

The Reeve shop technicians made house calls. On 19 August 1666 (less than two weeks before the Great Fire of London turned the lives of Londoners like Pepys upside-down), Reeve's son visited

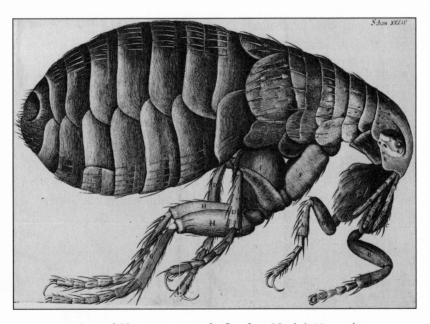

3. Large, fold-out engraving of a flea, from Hooke's *Micrographia*

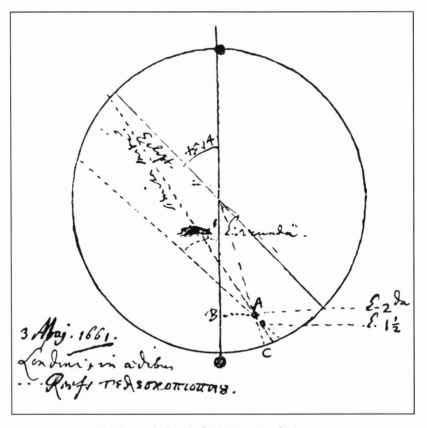

4. Huygens's sketch of 1661 transit of Mercury

Pepys by appointment (his father's professional career was brought abruptly to a close when he killed his wife 'by a knife flung out of his hand' in October 1664).[10] He brought him 'a frame with closes on, to see how the Rays of light do cut one another, and in a dark room with smoake, which is very pretty', and a magic lantern for projecting images on a wall 'very pretty'.[11] He stayed to dinner, and assisted Pepys with the telescope bought from him some months earlier: 'We did also at night see Jupiter and his girdle and Satellites very fine with my 12-foot glass, but could not Saturne, he being very dark.' Pepys expressed dissatisfaction that Richard Reeve junior lacked the theoretical competence of his father: 'It vexed me to understand no more from Reeve and his glasses touching the nature and reason of the several refractions of the several figured glasses, he understanding the acting part but not one bit of the theory, nor can make anybody understand it.'[12] Today, too, we would like the computer warehouse's technician to be able to solve all our software problems at the same time as installing the hardware and getting it running.

Hooke acted as an expert adviser to the Reeve shop, to satisfy just such demanding customers as Pepys. When Reeve delivered Pepys his overpriced microscope in 1664, he threw in a Hooke-designed scotoscope too (though he failed apparently to show Pepys how to use this 'curious curiosity' to advantage).[13]

It was in Reeve's shop that Hooke ran into an elderly Thomas Hobbes in the 1660s, in the process of choosing himself a telescope – and swearing profusely (Hobbes was renowned for his use of bad language). 'I found him to lard and seal every asseveration with a round oath, to undervalue all other men's opinions and judgments, to defend to the utmost what he asserted though never so absurd, to have a high conceit of his own abilities and performances, though never so absurd and pitiful,' reports Hooke. Hobbes was consistently cantankerously sceptical about the achievements of the new science, maintaining that there was no evidence that the advances in new technology were actually improving sense perception. In Reeve's shop, to Hooke's amusement, Hobbes 'would not be persuaded, but that a common spectacle-glass was as good an eye-glass for a thirty-six-foot glass [telescope] as the best in the world, and pretended to see better than all the rest, while holding his spectacle in his hand, which shook as fast one way as his head did the other; which I confess made me bite my tongue.'[14]

As far as Hobbes was concerned, the enthusiasm for scientific instruments was a fad that would pass. 'Every man that hath spare money' can acquire 'furnaces', 'telescopes' and 'engines'. 'They can get engines made, and apply them to the stars; but they are never the more philosophers for all this. It is laudable, I confess, to bestow money upon curious or useful delights; but that is none of the praises of a philosopher.'[15] It was a view he stuck to. As an uncompromisingly rationalist philosopher Hobbes was bound to deplore the way men like Hooke in practice fudged the boundaries between the predictions of abstract theory and the ingenious technical apparatus that produced useful results.

The Ingenious Mr Hooke

In the second half of the seventeenth century, Hooke's name crops up everywhere where new science was being undertaken and specialist equipment was required. He masterminded the technology behind a string of scientific 'discoveries' at the Royal Society. He could solve almost any technical problem involving scientific instruments at the drop of a hat. He provided London's clock- and instrument-makers with a stream of clever technical modifications to improve their products. He collaborated with Thomas Tompion on new kinds of clock balances and escapements, and with Reeve and Christopher Cock on superior lenses for telescopes and microscopes. After the death of Reeve senior in the 1670s, he shifted his custom, and his specialist advice, to yet another instrument-maker, John Cock, and collaborated with him in manufacturing quality reflecting telescopes for clients like the Royal Society in England, and Hevelius in Gdansk. He designed quadrants for the Greenwich Observatory, and self-levelling compasses for sea voyages. He engineered specific pieces of equipment (which the instrument-makers then built under his watchful eye) to test his own and others' theories of atmospheric pressure, motion, combustion, respiration. He dreamed of winning the handsome

5. Robert Boyle

cash prize on offer for solving the longitude problem, and several times claimed to have come up with plausible clock-based solutions. (The only scientific law credited to him today, however, involves the elasticity of springs. This was a by-product of his improvement in accuracy of the pocket-watch by substituting a coiled spring in its movement — an innovation on which, after considerable struggle, he failed to secure the patent.)[16]

Practical ingenuity such as Hooke possessed was essential for the success of the new experimental science. In the mid-1650s the aristocrat turned chemical enthusiast Robert Boyle retained the twenty-five-year-old Hooke on a generous salary to design customised experimental equipment, and to investigate experimentally any topic currently fascinating his employer. Effectively this meant that Boyle had acquired for his own personal use the *ad hoc* expertise Hooke had been providing in the shops of Reeve, Tompion and Cock. Hooke now oversaw the skilled manufacture of pieces of equipment specially designed for the Oxford laboratory where Boyle (at a discreet distance from the Cromwellian political regime in London) was engaged in his gentlemanly scientific investigations.

Just before the final collapse of the English Commonwealth, Boyle undertook a series of experiments to investigate the composition and 'spring' or compressibility of air — experiments in what today we would call 'pressure'. For this he needed a purpose-built piece of equipment that would enable him to control the amounts of air in a specified container. In early 1658 Boyle instructed Hooke and the London instrument-maker Ralph Greatorex to design and build for him a pneumatic pump — a closed vessel from which air could be extracted using a pump-mechanism, to produce something approaching a vacuum. Such a piece of equipment would allow Boyle to carry out experiments that depended on variable air pressure. The technical specifications for Boyle's pump included the fact that the chamber had to be accessible (unlike comparable air-pumps being built in France and the Netherlands) so that experiments measuring the effects of greater or lesser air pressure on living things (small birds, mice etc.) and on inanimate objects (barometers and clocks) could be carried out in its interior.[17]

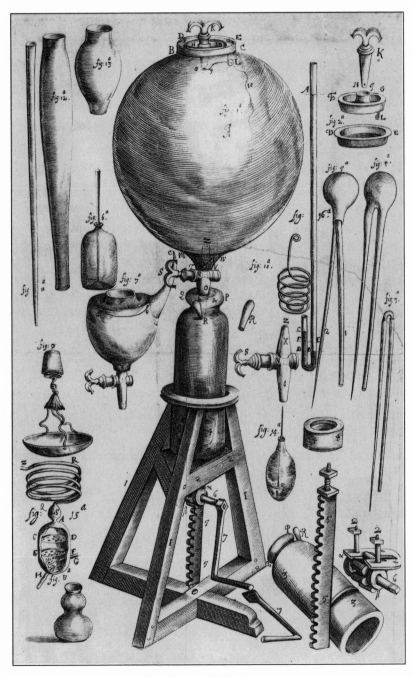

6. Boyle's remarkable air-pump

The impatient Hooke soon dropped Greatorex as co-designer of Boyle's air-pump, pronouncing the device he produced for testing 'too gross to perform any great matter'. By early 1659 Hooke had built an air-pump to his own specifications, devising customised components as he needed them, and supervising their manufacture by another London instrument-maker. The completed pump consisted of a large glass receiver (about thirty quarts in volume) with a four-inch opening at the top through which experimental apparatus could be inserted. A brass cylinder with milled valves, in which a sucker could be made to rise and fall by turning a handle attached to a geared ratchet, was mounted on a wooden frame below the receiver. A stop-cock was inserted at the bottom of the receiver and a valve at the top of the cylinder.[18]

The air-pump that Hooke built and repeatedly modified for Boyle became a celebrated feature of Royal Society meetings. Just two days after fellow member John Evelyn had attended Charles II's coronation at Westminster Abbey on 23 April 1661, he went with equal ceremony 'to the Society where were divers experiments in Mr. Boyle's Pneumatic Engine'.[19] Perhaps because of its association with the King's triumphant return, the air-pump was much in demand on occasions when Charles II himself or foreign dignitaries like the Danish and Genoese ambassadors visited the Royal Society. It was produced on the unique occasion known to us on which a woman was permitted to participate in the Society's investigations – when Margaret Cavendish, Duchess of Newcastle, visited in 1667 (according to Pepys she was 'full of admiration, all admiration').[20]

Pepys's own first taste of the Royal Society's activities, at the meeting he attended on the day he was proposed as a member by his Admiralty colleague Thomas Povey (Treasurer to the Lord High Admiral, the Duke of York) in February 1665, was a sequence of entertaining experiments on combustion making use of the famous air-pump. Hooke and Boyle were together in attendance to operate

the pump and give their explanation of events. 'It is a most acceptable thing,' wrote Pepys in his diary, 'to hear their discourses and see
their experiments, which was this day upon the nature of fire and
how it goes out in a place where the ayre is not free, and sooner out
where the ayre is exhausted; which they show by an engine on purpose.'[21] Yet the law discovered from experiments like these – that the
'spring' of the air is inversely proportional to the volume (PV is a
constant, where P is gas pressure and V its volume) is known today
as Boyle's Law.

The air-pump was temperamental; Hooke was the only person
who could reliably get it to work. The opening to the glass cylinder
had to be cemented shut each time it was used (various cements were
tried); air entered through the imperfect seal between the plunger
and its leather surround; the glass cylinder imploded or cracked.
In extended correspondence between Hooke and Christiaan
Huygens – the only other consistently successful operator of an air-
pump – both men were forced to admit that on numerous occasions
they had been unable to replicate a single successful experiment
because of some failure in the equipment

Huygens, who had been invited from the Hague to become a
founder member of the French Académie des Sciences in Paris in
1666, brought his own air-pump with him when he moved into
Louis XIV's royal library (the early home of the Académie). He
introduced the machine to his French colleagues with some simple
botanical experiments – evacuating the receiver and watching the
effect on plant germination and growth over a period of days.
Botany was a particular focus of interest among the members of the
Paris Académie. The air-pump produced far less spectacular effects
on plants (which simply failed to prosper) than the amusing experiments tried in London with animals (which collapsed and died).
Accordingly, Huygens's air-pump never achieved the symbolic cen-
trality in Paris of its counterpart in London.[22]

For the Royal Society in London it was important that their

elaborate performances for visitors worked. The public embarrassment caused by unsuccessful attempts at operating the air-pump
was of serious consequence. Unlike its French counterpart, the
Society depended entirely for support on private funding and the
voluntary participation of its gentlemen members. Its image was a
matter of the utmost importance. There was heated debate over the
Duchess of Newcastle's proposed visit – although she was a considerable scientist in her own right and had asked to attend out of
genuine interest, members were worried that a woman's presence at a
Society demonstration might lead to its being taken less seriously by
the discerning public. Pepys wrote in his diary that there had been
'much debate, *pro* and *con*, it seems many being against it, and we do
believe the town will be full of ballads of it'. In the end her elevated
rank settled the issue, and she was duly entertained with an air-
pump experiment.[23]

In November 1662, after a series of damaging fiascos when air-
pump demonstrations had to be abandoned as a result of faulty
components and incompetent demonstrators, the exasperated
Oldenburg prevailed upon the members to appoint Hooke to the
permanent paid post Curator of Experiments, a job tailored
expressly to his talents. (They first respectfully asked Boyle to release
Hooke from his duties in his laboratory to allow him to accept the
offer.) At least with Hooke in constant attendance the air-pump
stood some chance of functioning properly.

Hooke's bravura as an experimentalist – his pure showmanship –
stood the Royal Society in excellent stead over the years. If need be
he was prepared to experiment on himself. On one occasion in
1671 he devised a man-sized chamber for the air-pump, and vol-
unteered to occupy it while it was evacuated. Fortunately for Hooke,
the pump performed middlingly, emptying only about a quarter of
the air from the container. The sensations he reported when he
came out of his airless container were giddiness, deafness and pains
in the ears.[24]

Meanwhile, having lost Hooke's day-to-day services to the Royal Society, Boyle had to take on another technically minded laboratory assistant to carry on refining the air-pump's design. He chose a young Frenchman, Denis Papin, who had been assisting Christiaan Huygens with his related pneumatical machine in Paris. Boyle's 1680 published account of the experiments they conducted acknowledges the central nature of Papin's (and by implications Hooke's) scientific contribution. The air-pump was, Boyle writes, of Papin's own design, and differed materially from the one used before his arrival. 'I gave him the freedom to use his own [air-pump],' Boyle reported, 'because he knew how to ply it alone, and how to repair it more easily.' He also admitted in print that Papin had devised a number of the air-pump experiments himself ('as if they had been formed in his own brain'), and that Papin wrote up the experiments for publication (even though his English was far from perfect), because Boyle was suffering from a painful attack of the stone at the time: 'Being infirm in point of health, and besides, sur-rounded by many businesses, I was enforced to leave the choice of words to monsieur Papin.'[25]

Papin was a Huguenot, one of the many seventeenth-century sci-entists whose Calvinist faith forced him to leave France and thereby made his technical skills available in Protestant England, and later Germany. He is best remembered for his invention of the pressure cooker in 1679.

It was because of the prominence the Royal Society was giving to innovative experimental equipment like the air-pump, that the unknown young Isaac Newton chose to make his first bid for the attention of the London scientific community by presenting the Royal Society with an 'invention' (an improvement to a technical astronomical instrument). Just before Christmas 1671 Newton's senior Cambridge colleague Isaac Barrow brought to London a small reflecting telescope, under a foot long, which his protégé had made. The telescope used a highly polished spherical mirror in

7. Boyle's assistant Papin with his patented pressure-cooker

place of the usual second lens of the refracting telescope, bending the light beam, and thus making it possible to produce a far more portable instrument than the increasingly huge (50- to 60-foot-long) refractors.[26]

The telescope caused quite as much of a stir as Newton had hoped for, especially in view of its obvious importance as a device for making astronomical observations at sea (because its size made it so portable). It gained enthusiastic endorsement from the Royal Society President, Viscount Brouncker, who as Commissioner and Assistant Comptroller to the Navy Board immediately recognised its strategic potential. In early January 1672 Secretary Oldenburg responded, urging Newton to allow the Society to register an international patent for the instrument on his behalf:

> Your Ingenuity is the occasion of this addresse by a hand
> unknowne to you. You have been so generous, as to impart to
> the Philosophers here, your Invention of contracting
> Telescopes. It having been considered, and examined here by
> some of ye most eminent in Optical Science and practise, and
> applauded by them, they think it necessary to use some meanes
> to secure this Invention from ye Usurpation of forreiners.[27]

The telescope, however, had merely been Newton's ploy for getting the Royal Society's attention. He was far more interested in the reception of a mathematical paper on optics which he now sent to Oldenburg, containing 'an accompt of a Philosophicall discovery wch induced mee to the making of the said Telescope, & wch I doubt not but will prove much more gratefull then the communication of that instrument'.[28]

But because Newton's original submission had been a piece of technology, the Society handed over the task of formal assessment to their in-house technical expert, Hooke, rather than to one of their distinguished pure mathematicians, like Wren. Flustered by the practical implications of the telescope (which was an improvement on anything he had himself designed up until then), Hooke responded with a devastating critique of Newton's theoretical work. Blithely unaware of the entirely original mathematics on which Newton's theory of refracted light was based, he announced crudely, and somewhat spitefully, that its postulates was insufficiently experimentally validated. Newton withdrew his work on optics from

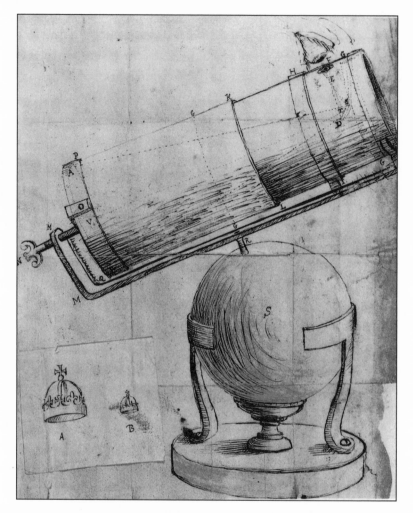

8. Royal Society sketch of Newton's reflecting telescope

publication, and returned to isolation at Trinity College, only emerging to become a fully participating figure in the Royal Society (and finally publishing his ground-breaking *Optics*) in 1704, after Hooke's death.

There is a certain irony in the fact that although Newton withdrew his theory of the refraction of light in the face of Hooke's disparaging comments, he was far less inclined to let go of his priority claim to the compact reflecting telescope. He continued obstinately to back his telescope against Hooke's 'improved' version. 'It is not for one man to prescribe Rules to the studies of another,' he responded furiously to Oldenburg, after he received Hooke's comments on 'perfecting' refracting telescopes.[29] Through challenging Newton head-on, Hooke actually encouraged Newton to take a continuing active interest in telescope-manufacture (Newton personally oversaw the subsequent manufacture of parts for an improved reflecting-telescope in Cock's shop, and to some extent displaced Hooke there as technical adviser).

Those who knew Newton at the time recalled his pride at his mastery of the technical side of instrument-making: 'I asked him,' a colleague later recalled, 'where he had [his telescope] made, he said he made it himself and when I asked him where he got his tools said he made them himself and laughing added if I had staid for other people to make my tools and things for me, I had never made anything of it.'[30] But the incident permanently soured Newton against those whom he considered to be 'mere' instrument-designers, with only a limited grasp of the theoretical premises on which their instruments were based: 'A Vulgar Mechanick can practice what he has been taught or seen done, but if he is in an error he knows not how to find it out and correct it, and if you put him out of his road, he is at a stand.'[31]

Hooke was certainly at his worst, temperamentally, when a technical challenge from the likes of Newton stung him into competitive retaliation. On the other hand, when his whole attention was focused on his microscopic observations, unravelling nature's minute mysteries, and transferring his revelations to the printed page, he was engagingly modest. 'I here present the world my imperfect indeavours,' he announced in the preface to *Micrographia*. His drawings of the minute structures of natural bodies would, he hoped, enable men 'to discern all the secret workings of Nature, almost in the same manner as we do those that are the productions of Art, and are managed by wheels, and engines, and springs, that were devised by human wit.' All that it took for such fundamental insights was 'a sincere hand and a faithful eye, to examine, and to record, the things themselves as they appear.'[32]

Hooke's keen eye for observation and his steady, accurate hand were those of the trained draughtsman with a flair for drawing (anecdotally he always maintained that he would have become a painter if the smells of painter's solvents had not made him nauseous as a boy). He shared this training in technical drawing with his close London colleague Christopher Wren. It was Wren, as Hooke was anxious to point out, who had first seen the potential to entertain the scientific amateur by making meticulous drawings of the 'small animals' seen under the microscope. As early as 1661 Wren had diverted Charles II with drawings of a louse, a flea and the wing of a fly. Subsequently the King asked the Royal Society if Wren could make him further such drawings, along with a lunar globe. The drawings and globe (ten inches in diameter, made of pasteboard, painted and carved in relief to show the moon's 'seas', mountains and craters) were received 'with particular Satisfaction, and ordered by His Majesty to be placed among the Curiosities in his Cabinet'. Christiaan Huygens saw them there in 1663, as did a number of other visiting foreign dignitaries.[33]

The task of making extensive further microscope-based drawings at His Majesty's command was, however, passed to Hooke on the grounds that Wren was too busy to complete the task. On 13 August 1661 the senior courtier Sir Robert Moray told Wren that 'In Compliance with your Desire to be eased of the further Task of drawing the Figures of small Insects by the Help of the Microscope, we have persuaded Mr. Hooke, to undertake the same Thing.'[34]

Once Hooke became Curator of Experiments, producing one or more such drawings for each of the Royal Society's regular meetings became part of his job. On 15 April 1663, for instance, he:

shewed two microscopial schemes, one representing the pores
of cork, cut both traverse and perpendicular; the other a
Kettering-stone, appearing to be composed of globules, and
these hollow ones, each having three coats sticking to one
another, and so making up one intire firm stone. He was
desired to examine the barks of other trees, and to write down
all that he should observe about these and the like appearances;
and also to bring in to the next meeting the representation of
the little fishes swimming in vinegar.[35]

Micrographia is thus in part the Curator's record of the sorts of
things that had entertained the fellows of the Royal Society and its
royal patron; the reader engages in new science in the company of
the Society's experimental impresario and practical wizard, Hooke.

Micrographia is a collaborative scientific venture several times over.
Like all Hooke's published works, it assembles (rather unsystemat-
ically) materials drawn directly from the day-to-day activities of
the Royal Society, and the experimental services Hooke rendered
individual members as its Curator. And it is enduring concrete testimony to an important and lasting scientific partnership between
Robert Hooke and Christopher Wren, just as the air-pump is
enduring evidence of Hooke's scientific collaboration with Boyle.
For when Wren's microscope-based drafting project was handed on
to that ever-resourceful facilitator, Hooke, he did not relinquish
it — he was simply relieved of the tedious work involved in completing the project. *Micrographia* not only includes drawings of
precisely the entertaining 'small animals' Wren had already produced for the King, it also contains an elegantly executed drawing of
the magnified surface of the moon, complete with mountains and
craters, of the kind Wren had undertaken, and which had formed
the basis for Wren's three-dimensional lunar globe.

The practical project of making a larger-scale lunar globe for the
Royal Society, to which Wren had committed himself, was likewise
handed on to others. In January 1667 Wren was reminded that the

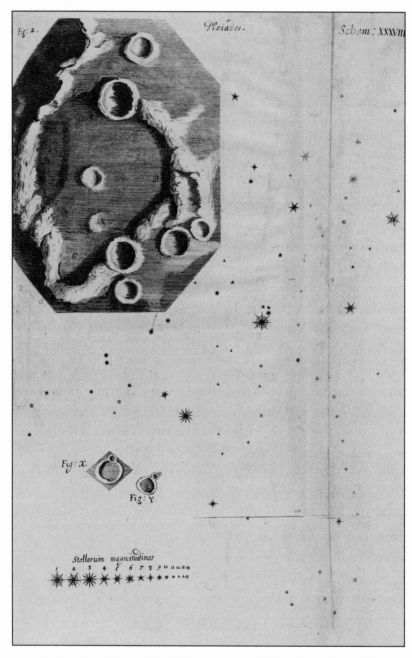

9. Engraving of the surface of the moon, possibly based on a drawing by
Wren

Society had been waiting over three years for 'the telescopical moon formerly promised by him'. A plan was devised for making a globe with the minimum of effort on Wren's part. The King would be asked to lend his globe, and Wren would employ a workman to scale up and copy the features on to a larger version.

Monumental instruments

In spite of Hooke's generous acknowledgement of Wren in the preface to his *Micrographia*, the most memorable partnership between the two men was not that entertaining volume, but rather one forged around a project on a far grander scale — one that was to make a permanent impact upon London's skyline.

On 25 August 1664 Hooke wrote to his former employer Robert Boyle, in Oxford, to keep him informed of some scientific experiments he had carried out for the Royal Society 'on the top of St. Paul's steeple' a few days earlier. The experiments were simple ones, because on this, Hooke's first trip up the semi-derelict ancient building (whose dramatic spire had been struck by lightning a century earlier), he had brought little equipment with him:

> One was, that a pendulum of the length of one hundred and eighty foot did perform each single vibration in no less time than six whole seconds; so that in a turn and return of the pendulum, the half-second pendulum was several times observed to give twenty-four strokes or vibrations. Another was, that this long pendulum would sometimes vibrate strangely, which was thus: The greatest part of the line, by guess about six score foot of the upper part of it, would hang directly perpendicular, and only the lower part vibrate; at what time the vibrations would be much quicker, and this though there was a weight of lead hung at the end of the string of above four pounds weight.

While he was up there, he did a bit of surveying: 'In another place of the Tower, where I had very clear perpendicular descent, I with the plumb-line found the perpendicular height of it two hundred and four feet very near, which is about sixty feet higher than it was usually reported to be.' He told Boyle that he would return to the same location to 'try the velocity of descent of falling bodies, the Torricellian [barometer] experiment, and several experiments about pendulums and weighing'.[36]

Two weeks later Hooke wrote to Boyle again, with an account of the experiments he had conducted as promised. They required a good measure of on-the-spot improvisation, because of the precarious structural state of the steeple. To begin with, there were no suitable cross-beams from which to suspend Hooke's equipment:

> The steeple being without any kind of lofts, but having only here and there some rotten pieces of timber lying across it, I caused a rope to be stretched quite cross the top, and fastened; in the midst of which I fixed a pulley, through which I let down the string and weight to the bottom (for only in the very middle of the steeple was there a broad clear passage from top to bottom) and to this I could not at the top approach within eighteen foot. Having thus let down the rope, those that were at the bottom hung on this mercurial tube [the barometer] (which I had already marked and stopped, and set ready before I went up) a large weather glass (which moved by the rarefaction and condensation of the air only, which I had likewise marked and stopped) and a sealed thermometer, which I had likewise marked.

In spite of Hooke's careful precautions, things did not go entirely according to plan:

> After these [instruments] were drawn up, and by the contrivance of another pulley I had drawn them to me, I found the thermometer, the glass being but thin, broken. The quicksilver, upon opening the cock, I found to fall very

considerably, which since, upon measuring, I find 25/48 of an inch. The weather-glass I found to be risen somewhat more than two inches. Then closing them again, I caused them to be let down; and giving them charge not to let it quite down till I called to them from below, I went down myself, and found, upon opening the mercurial tube, that it rose exactly to its first station; as did also the weather-glass.

I had designed to have tried many others [other experiments] then, but the night came so fast, that I could hardly see to get up again, and give order for the clearing of the lines.[37]

Still, Hooke decided this was a good location for experiments whose execution required substantial differences in elevation. A week later again, he reported to Boyle that he had repeated the 'Torricellian experiment', and carried out more pendulum experiments at Old St Paul's, this time with the President of the Royal Society (Lord Brouncker) and Sir Robert Moray in attendance (presumably safely at the bottom of the structurally unsafe steeple). Boyle, in the meantime, wrote back to the Royal Society secretary Oldenburg, grandly suggesting various modifications to Hooke's experiments, and proposing some experiments of his own with magnetism, which might also be tried at the bottom and top of St Paul's steeple – 'which experiments', the records of the Royal Society tell us tactfully, 'though esteemed considerable, yet were thought not practicable at the said steeple, by reason of the iron there mixed every where'.[38]

Hooke chose St Paul's as a convenient 'height' upon which to conduct his experiments because he was already engaged in work there, surveying the spire to assess its structural state.[39] One of Hooke's several professional occupations was that of surveyor. In 1663 a government commission was appointed to oversee the restoration of Old St Paul's. The building was pronounced unsafe, and it was proposed that the damaged nave and the steeple, which had been struck by lightning in the sixteenth century, be demolished immediately. Christopher Wren was called in to give a second opinion.

Wren, a considerable mathematician, had been Professor of Astronomy at Oxford since autumn 1661, but on top of his academic brilliance he had already established his credentials as a structural surveyor and architect of the Sheldonian Theatre at Oxford. Wren's design for the roof of the Sheldonian, which spanned a hitherto impossible space without supporting pillars, was regarded as an engineering marvel. The model for the building was exhibited at a meeting of the Royal Society, at which Wren was asked to provide a written description of the roof-structure, 'to remain as a memorial among the archives of the society'. In his *Natural History of Oxfordshire* (1677), Robert Plot, the first curator of the Ashmolean Museum, included the Sheldonian as one of the 'remarkable Curiosities' of the county.[40]

The 'restoration' of St Paul's was a matter of some sentimental importance to the restored monarchy of Charles II. The great neo-classicising architect and designer Inigo Jones had partially completed a magnificent classicising face-lift for the cathedral for Charles I, refacing the exterior of the nave, and building a classical façade with pillars, when civil war brought the project to a standstill. During the Commonwealth period, Oliver Cromwell – for whom Inigo Jones's refurbishment was far too Roman and papist – considered turning the 'Crosse building of St. Paul's' into a fortified base for himself 'to bridle the City of London'; Jonas Moore (not yet knighted) reputedly constructed a feasibility model.[41]

In May 1666 Wren produced a written report, accompanied by preliminary designs, in which he proposed encasing and refacing the entire interior of the nave of St Paul's (as Inigo Jones already had done the exterior thirty years earlier) and enclosing the 'tower's world of Scaffolding poles' (from which presumably Hooke had hung his pulleys) with 'a spacious Dome or Rotunda, with a Cupola or hemispherical Lantern with a Spiring Top' rising 'proportionately' above it.[42]

In his diary entry for 27 August 1666, John Evelyn (amateur of

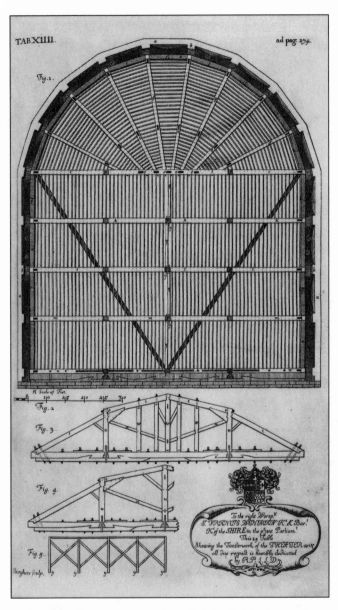

10. Plan and section of Wren's roof of the Sheldonian Theatre

classical architecture) records an on-site visit to St Paul's, following on from Wren's report:

> I went to St. Paules Church in London, where with Dr. Wren, Mr. Pratt, Mr. May, Mr. Thomas Chichley, Mr. Slingsby, the Bishop of London, the Dean of St. Paul's, and several expert workmen, we went about to survey the general decays of that ancient and venerable church, and to set down the particulars in writing, what was fit to be done, with the charge thereof: giving our opinion from article to article. When we came to the Steeple, it was deliberated whether it were not well enough to repair it only upon its old foundation, with reservation to the 4 Pillars: this Mr. Chichley and Pratt were for; but we totaly rejected it and persisted that it requird a new foundation, not only in regard of the necessity, but for that the shape of what stood was very mean, and we had a mind to build it with a noble Cupola, a form of church building, not as yet known in England, but of wonderfull grace: for this purpose we offerd to bring in a draught and estimate, which (after much contest) was at last assented to, and that we should nominate a Committee of able workmen to examine the present foundation.[43]

Five days later Evelyn's scrupulous structural survey and Wren's report had both ceased to have any relevance. During the night of 2 September 1666 and the five days that followed, the Great Fire of London entirely gutted Old St Paul's, along with about seven-eighths of the City of London (395 acres): 13,200 houses, 87 churches, 44 livery halls, six prisons, four bridges, the Customs House, Guildhall and Exchange were destroyed, along with goods estimated at £3.2 million.[44] On his next visit to the site Evelyn describes, in a state of shock, how the searing heat had cracked Inigo Jones's façade, and 'that beautiful portico now rent in pieces, flakes of vast stone split in sunder, and nothing remaining entire but the inscription in the architrave which shewing by whom it was built, had not one letter of it defaced'. 'Thus lay in ashes that most venerable church.'[45]

11. Wyck's drawing of Old St Paul's following the Great Fire

By 1670, when Wren was made London's chief surveyor by Charles II himself, and put in charge of the major rebuilding effort necessitated by the Great Fire, Hooke was working for him on a regular basis (at a basic salary of £50 per annum), and acting in what architects today would regard as an engineering capacity, as well as as surveyor. Hooke was in any case part of the complementary team appointed for the project at the same time by the Corporation of the City of London (strictly, the client). Thus, Wren was London's leading architect and planner; Hooke the leading surveyor and engineer acting on behalf of the client (the Corporation of the City of London). On 12 November 1673, Wren was appointed by Royal Warrant as architect of the new St Paul's: two years and three designs later, building actually began, in the summer of 1675.[46]

A horizon-dominating, spectacular St Paul's, rising phoenix-like from its own ashes, was symbolically central to London's rebuilding

effort, and St Paul's-related work occupied a significant portion of
the Wren architectural office's time from the early 1670s until the
building was eventually completed in 1710–11. Hooke was closely
involved. Between August 1672 and December 1680 alone[47] there
are more than a hundred entries in Hooke's diary indicating a visit
to the St Paul's site, sometimes accompanied by Wren, sometimes
missing Wren but meeting one or more of the other members of
the Wren office there.[48] Hooke instructed bricklayers; he discussed
the quality of marble.[49] Later, when advising Lord Conway, for
whom he designed Ragley Hall, Hooke warned that the mortar
and brickwork should be completed well before winter set in to
allow the mortar to dry thoroughly, otherwise it would only have to
be taken down and rebuilt, 'as I have found in the building of St
Paules'.[50]

Shortly after Wren's appointment to the post of chief surveyor,
Hooke made him a model out of chain links forming an inverted
dome, held in shape tensionally by the force of gravity. He added
weights and additional links until he had achieved the bell-like
shape he desired. He proposed that if the shape were imagined
dipped in plaster and turned bowl uppermost, the tension forces
reverse into pure compressions, with a distribution that perfectly
simulates a completed masonry-constructed dome with buttresses
(inverting the added weights into pressures) at critical points. The
issue is one of converting pure geometry into mechanics – trans-
forming the conceptual form on the page into built form.

Hooke alerted the Royal Society to his insight about arch-
construction several times in late 1670; Wren concurred, and they
jointly made verbal presentations at a Royal Society meeting on 19
January 1671. A year after his original announcement, on 7
December 1671, Hooke 'produced the representation of the figure
of the arch of a cupola for the sustaining such and such determinate
weights, and found it to be a cubico-paraboloid conoid [actually it
is no such thing]; adding that by this figure might be determined all

12. Wren's Great Model of St Paul's

the difficulties in architecture about arches and butments'.[51] On Saturday, 5 June 1675, Hooke records in his diary: 'At Sir Chr[istopher] Wren. At Mr. Boyles. At Mr. Montacues with Fitch. . . . Told Montacue of Pillers 20 foot high for £10 [altogether]. Told Sir Ch[ristopher]Wren of it. He promised Fitch at Paules. He was making up of my principle about arches and altered his module by it.'

Hooke recorded his rationale for such a masonry form based on the theoretical inversion of a hanging catenary in an anagram inserted at the end of the 1676 printed version of his 1674 Cutlerian lectures. The announcement runs: 'The true Mathematical and Mechanichal [*sic*] form of all manner of Arches for Building, with the true butment necessary to each of them. A Problem which no Architectonick Writer hath ever yet attempted, much less performed', followed by the curious anagram.[52] Decoded the formula ran: 'As it hangs in a continuous flexible form, so it will stand contiguously rigid when inverted.'[53]

Hooke's solution to the technical problems posed by Wren's ambitious design for the dome of St Paul's showed significant engineering acumen. The inner dome and its ring of inward-sloping pillars do appear to conform to a self-supporting inverted catenary. The masonry cone above it, its buttresses concealed by the peristyle that

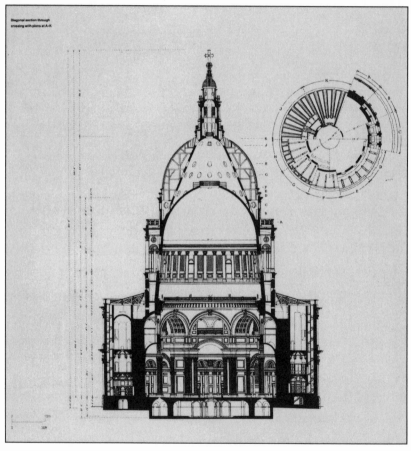

13. Cross-section of measured drawing of the dome of St Paul's

14. Isometric drawing of the dome of St Paul's

supports the 70-foot lantern, weighing a formidable 850 tons, also displays the influence of these catenary arch discussions.

The unique construction suggests a further connection between Hooke's scientific activities and his engineering. The cone has a small 'oculus' or light-admitting aperture at its top; below this is the inner dome, with a considerably larger oculus. The larger aperture gives the illusion that more light is entering St Paul's than is in fact the case – an arrangement that resembled the two apertures of a microscope with one lens stopped down to reduce the coloured fringes of chromatic aberration. To the observer with his microscope, the reduced aperture does not alter the illusion that light is filling the eyepiece lens; similarly, the person standing beneath the dome of St Paul's believes the lower aperture to be 'filled' with light.

The Great Fire that gutted Old St Paul's destroyed one of Hooke's carefully chosen locations for pendulum and barometer experiments. But it did not destroy his belief that buildings on a grand scale could be conceived of as multi-purpose – part monumental building, part scientific instrument or oversized piece of scientific equipment. If we look with a seventeenth-century scientist's eye at other post-Fire buildings, we can recognise designs with the possibility of double function in mind. In their designs for post-Fire monumental buildings in London, Wren and Hooke, close friends, both enthusiastic astronomical observers, both prominent members of the Royal Society, retained a commitment to the view that large man-made monuments offered tremendous possibilities as purpose-built 'heights' for scientific experiments.[54]

Among the many London building projects on which the two men worked jointly was the Monument to the Great Fire itself. It is now generally agreed that it was Hooke's design for the Monument that was finally built (his diary records his putting the finishing touches to the scale model for the 'Piller' in October 1673, and a final drawing in his hand is endorsed by Wren). According to a contemporary report, Wren and Hooke chose the form for the

1. Newton in later life

EDMVND. HALLEIVS LL.D.
GEOM. PROF. SAVIL. & R.S. SECRET.

2. Edmond Halley

3. Robert Boyle

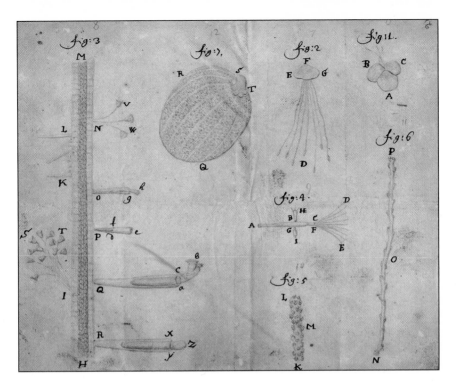

4. Original red chalk drawing of duckweed by Leeuwenhoek

5. Vermeer's *View of Delft*

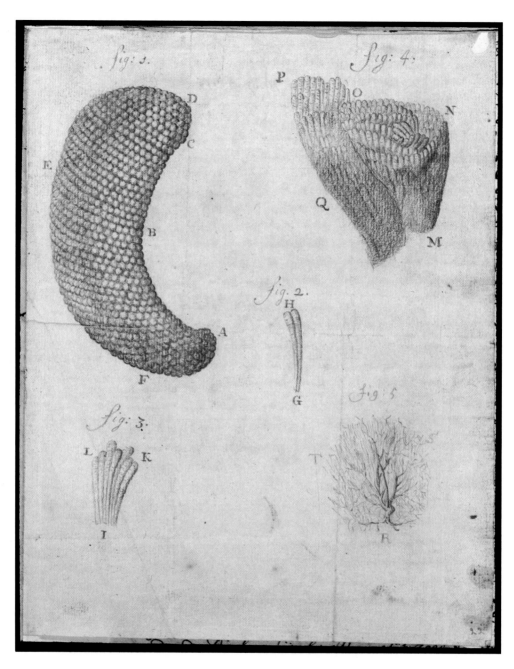

6. Original Leeuwenhoek drawing of a bee's organs

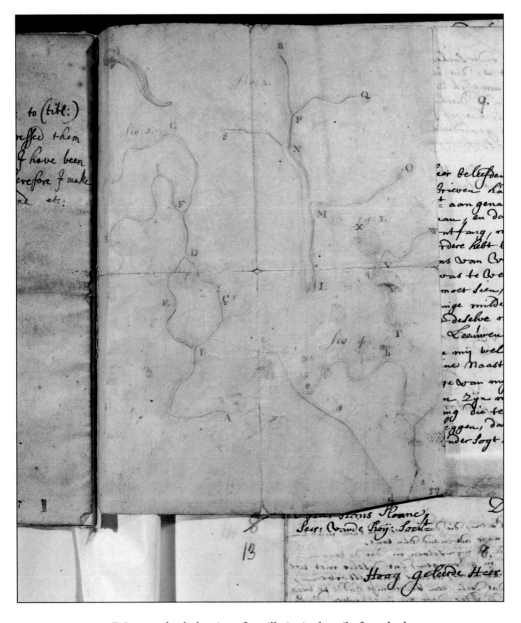

7. Leeuwenhoek drawing of capillaries in the tail of a tadpole

8. Dutch seventeenth-century flower painting with insects

9. *Perspective from a Threshold* by van Hoogstraten, which Pepys saw at
Thomas Povey's house

10. Van Hoogstraten's perspective box

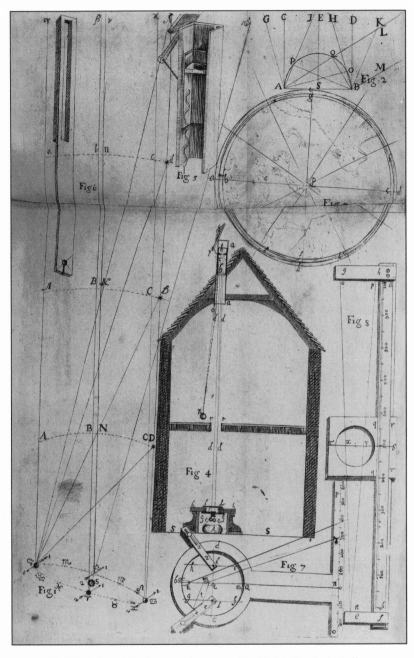

15. Hooke's zenith telescope, constructed inside his Gresham College
home (fig. 4)

202-foot-high Monument – a hollow pillar – in the hope that it would be possible to mount an immense zenith telescope inside, to attempt to observe stellar parallax (like the 'tubeless' telescope Hooke had already constructed through his house at Gresham College, which proved insufficiently stable for reliable measurements). The Monument too proved unsuitable.[55]

For the stellar parallax experiment, a selected fixed star had to be observed in the same night sky position, precisely six months apart, and any minute displacement in its position recorded; errors introduced by altered conditions in the observing equipment (caused by meteorological factors, etc.) made detection of such hypothesised small alterations impossible. The Monument did, however, prove a suitable location for more modest kinds of experiments: on 16 May 1678 Hooke recorded in his diary: 'At Fish Street pillar [Monument] tried mercury barometer experiment. It descended at the top about ⅓ of an inch.'[56] On 23 May he 'directed experiment at Column. Lent Mr. Hunt a cylinder to do it.'[57] The Proceedings of the Royal Society for 30 May record that Hooke measured the pressure at various stages as he came down the Monument's steps, but that he was not entirely happy with the accuracy of his equipment:

> he had observed the quicksilver to ascend by degrees, as near as
> he could perceive, proportional to the spaces descended in
> going down from the top of the column to the bottom: but
> because the said stations of the mercury were different from
> one another but very little, and so it was not easy to determine
> the certain proportions of the one to the other; therefore he
> proposed against the next meeting an experiment be tried at the
> same place with an instrument which would determine that
> distance an hundred times more exactly: which instrument also
> he there produced, in order to explain the manner thereof, it
> being made upon the same principle with the wheel barometer,
> but more curiously wrought.[58]

16. The Monument to the Great Fire, designed collaboratively by
Hooke and Wren

In December 1678 Hooke measured the height of the Monument.[59] Perhaps he hoped that it would match the height he had previously found the tower of Old St Paul's to be, and that his Torricellian and pendulum experiments could continue the series he had begun twelve years earlier at the top and bottom of Old St Paul's tower. The inscription on the base informs us that the height of the Monument was significant – the distance from it to the baker's shop where the Great Fire had started. Since Wren and Hooke chose the site, this tells us only that the Monument was designed to rise to a height of somewhere around 200 feet. One way or another, the Monument was a memorial to London's curious intellects, and that mental moment which immediately preceded the Great Fire.

Wren, too, could see the attraction of designing expensive, large-scale public buildings so that they could double as astronomical instruments. Several of the buildings designed collaboratively with Hooke for scientific purposes incorporate bricks and mortar structural features to facilitate scientific activities. The Royal Observatory Building at Greenwich was the only such edifice actually built (and even so, the haste with which Sir Jonas Moore insisted on its being completed prevented their incorporating a purpose-built meridian-line wall from the outset). The 1667–8 plans for a proposed new-build home for the Royal Society in the grounds of Arundel House (never actually executed), however, included a 'passage gallery' running the whole length of the attic storey, which was designated 'for tryal of glasses & other experiments that require length'.[60]

In February 1704 (by which time Hooke was dead), Wren proposed using the massive, still incomplete structure of St Paul's Cathedral itself as a super-sized zenith telescope.[61] Constantijn Huygens, Christiaan's elder brother, had given the Royal Society the lenses of a 123-foot telescope in 1692 'with the Apparatus for using them without a Tube'. Wren's plan was to mount the telescope

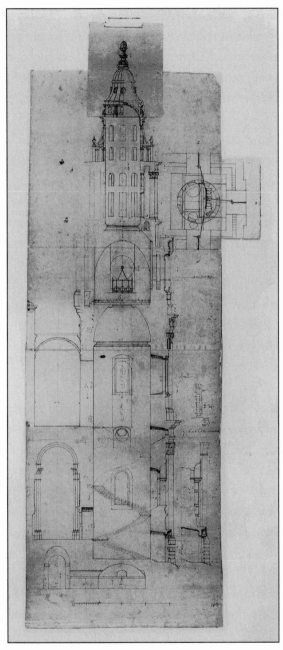

17. Drawing in Wren's hand of the south-west tower of St Paul's

in the south-west staircase of St Paul's, and Hodgson, who on Wren's recommendation had been employed and trained by Flamsteed at the Royal Observatory (where zenith observations had also been attempted), was to have made the observations. Again the plan failed: Huygens's telescope was, in fact, too long.[62]

The challenge Christopher Wren faced when given the task of designing the post-Fire St Paul's Cathedral tends today to be seen in terms of the relationship between it and the antique, classicising influences on architectural design newly introduced into Europe by continental scholars of Latin and Greek, who were vigorously uncovering and publishing in accessible form texts on Roman and Greek architecture long hidden from public view: Vitruvius's comprehensive ancient handbook on building techniques and styles, Alberti's treatise on buildings, Palladio and Serlio.

It is true that both Wren and Hooke bought and read such books (their titles crop up in Hooke's diary, and are recorded in the printed catalogues of sale for the two men's libraries, auctioned after their deaths). However, Wren and Hooke each owned a copy of William Dugdale's *History of [Old] St. Paul's cathedral in London, from its foundation untill these times*. This illustrated volume, originally published in 1658, before St Paul's was destroyed, was reissued after the Restoration. Its magnificent engravings of the intact cathedral became a symbol for the revival of the grandeur of monarchy, and shaped Wren's aspiration to rebuild a monument of equivalent significance. Furthermore, Wren's practical involvement with St Paul's predated the fire by a number of years, and was bound up with that other set of activities in which Wren and Hooke were engaged together: the experimental science of the early Royal Society.

The rebuilding of St Paul's took over forty years. The partially completed cathedral became the focus for Londoners' dreams of all kinds, including the idea that it might house London's first public library. On 13 February 1684 John Evelyn wrote in his diary:

Dr. Tenison communicating to me his intention of Erecting a
Library in St. Martines parish, for the publique use, desird my
assistance with Sir Christopher Wren about the placing &
structure thereof: a worthy & laudable designe: He told me
there were 30 or 40 Young Men in Orders in his Parish, either,
Governors to young Gent: or Chaplains to Noble-men, who
being reprov'd by him upon occasion, for frequenting Taverns
or Coffè-houses, told him, they would study & employ their
time better, if they had books: this put the pious Doctor upon
this designe, which I could not but approve of, & indeede a
greate reproch it is, that so great a City as London should have
never a publique Library becoming it: There ought to be one at
St. Paules, the West end of that Church, (if ever finish'd),
would be a convenient place.[63]

In the event, access to the completed Cathedral's library was more
conventionally restricted.

Engravers and instruments

The original charter of the 'Royal Society, for the improvement of
naturall knowledge by Experiment', passed on 13 August 1662,
laid down that it was 'to Consist of a President, Council, Fellows,
Secretaries, Curators, Operator, Printer, Graver & other officers'.[64]
The charter thus acknowledged the key role of printers and
engravers – those who produced the publications and visual images
on which the Society's reputation largely depended for those outside
the élite inner circle. The Society's official *Philosophical Transactions*, in
which all its important scientific communications were circulated
and preserved for posterity, have indeed largely shaped our view of
its activities (and we should remember that such official publications
never record botched or failed experiments). The engraved plates
were just as influential.

For early scientific enthusiasts, the beauty of the interior structure of a dome, or the patterns of nerves in the human neck, or the detail of the viscera of a bee, were equally compelling; the challenge, to convey that beauty to the ordinary observer. Both Wren and Hooke worked on inventions to improve the engraving process itself, to make yet better plates for their publications. When the London public became impatient with the time it was taking to rebuild St Paul's and threatened to sack the architect, Wren released a series of beautifully executed engravings purporting to show the completed cathedral, in the hope that these would satisfy them (there is, in fact, little resemblance between the dome as completed and the imagined versions lovingly depicted in the engravings).

John Wilkins, Bishop of Chester, responding to the beauty of his protégé Robert Hooke's engraved plates in *Micrographia*, observed how crude even the finest engraving appeared alongside the beauty of nature, when both were seen through the microscope:

> Whatever is Natural doth by that appear, adorned with all imaginable Elegance and Beauty. There are such inimitable gildings and embroideries in the smallest seeds of Plants, but especially in the parts of Animals. In the head or eye of a small Fly: such accurate order and symmetry in the frame of the most minute creatures, a Lowse or a Mite, as no man were able to conceive without seeing of them. Whereas the most curious works of Art, the most accurate engravings or embossments, seem such rude bungling deformed works as if they had been done with a Mattock or a Trowel.[65]

In the same 1664 letter in which Hooke told Boyle about the pendulum and barometer experiments he had conducted at the top of Old St Paul's, he asked his former employer for some further instructions concerning a portrait-engraving (commissioned as the frontispiece of a volume of Boyle's published works) that the eminent engraver William Faithorne was making of him:

Mr Faithorne has promised me to make all possible speed with that you ordered him, but he does desire a little farther directions. Whilst I was writing this, Mr Faithorne has sent me the sketch, which I have enclosed, to see whether you will have any books, or mathematical, or chemical instruments, or such like, inserted in the corners, without the oval frame or what other alteration or additions you desire.[66]

In the next letter (the one reporting the broken thermometer), Hooke reported that he himself had:

made a little sketch, which represents your first [pneumatical] engine placed on a table, at some distance beyond the picture, which is discovered upon drawing a curtain. Now, if you think fit, I think it might be proper also to add, either by that, or in the corners, your last emendation of the pneumatick engine. One word or two I beseech you of directions in this particular.[67]

For the purposes of the memorial engraving, Hooke represents the air-pump as belonging exclusively to Boyle; and, in Boyle's engraved portrait, this much-publicised and much-celebrated piece of equipment symbolises Boyle's pre-eminent position within the new science.[68] It is Hooke who suggests the introduction of the pneumatic engine (the first pull of Faithorne's portrait shows a simple landscape). It is an intelligent suggestion in terms of public relations for the Royal Society and its activities. Large, readily recognisable, expensive, producing dramatic experimental effects and, above all, nail-bitingly likely to go wrong, the pneumatic engine was to 'air' (gases) what the van de Graaff machine was for early experimentation with electricity. Even an audience with no understanding of science could be thrilled by the spectacle of its operation.

But Hooke's activities in and around the London print-shops were not confined to supervising the engraving and reproduction of

Guilhō Faithorne ad viu: delin: et sculp:

CROBERTVS BOYLE ARM:

18. Faithorne's engraving of Boyle and his air-pump

the frontispiece to Boyle's book. We discover from a letter written on 24 November 1664 that Hooke was responsible for seeing the entire Boyle volume (*New Experiments and Observations touching Cold* [1665]) through the press. Specifically, he was collating engraved illustrations with Boyle's written text – illustrations that he, Hooke, had personally identified the need for and drawn (and he was having trouble getting his engravers to produce their work on time):

> I have procured out of Mr. Oldenburg's hands some of the first sheets of your book on coldness; and I shall delineate as many of the [scientific] instruments you mention, as I shall find convenient, or (if it be not too great a trouble to you) as you shall please to direct. I think it will be requisite also, because your descriptions will not refer to the particular figures and parts of them by the help of letters; that therefore it would not be amiss, if I add two or three words of explication of each figure, much after the same manner, as the affections of the prism are noted in your book of Colours. The figures I think need not be large, and therefore it will be best to put them into one copper plate; and so to print them, that they may be folded into, or displayed out of the book, as occasion serves.[69]

While Hooke was explaining his choice of illustration, he remembered the other engraving job that he was in charge of in his old employer's absence:

> This puts me in mind to acquaint you also, that Mr. Faithorne has now at last promised me with all the asseverations imaginable, that he will not fail to finish your picture by the middle of next week at furthest, and therefore I think I shall employ Mr. Logan (who is an excellent graver also) [for the plates for your book] that I may not take Mr. Faithorne off from finishing that plate.[70]

Which in its turn reminded him to tell Boyle that the plates for his own *Micrographia*, unlike Boyle's portrait, were ready (apparently sitting in Faithorne's shop). Here it was the Royal Society's vetting of Hooke's text that was holding up publication:

> As for my own microscopical observations, they have been printed off above this month; and the stay that has retarded the publishing of them has been the examination of them by the several members of the Society; and the preface, which will be large, and has been stayed very long in the hands of some who were to read it. I hope I shall prevail with the printer to dispatch some time this or the next week.[71]

A month later, Hooke was still waiting for Faithorne to finish Boyle's portrait-print. Boyle was apparently not helping, but adding last-minute material to his volume. There is also a hint that Boyle had complained at not having Faithorne to execute the plates for his book:

> The plate for your book was graven before I received your last of Mr. Evelyn's. I have only taken notice of seven instruments, which you in those sheets I looked on have described; and those I so put into one small plate, that they will fold out of the book, when there is occasion. This last of Mr. Evelyn I have given a small draught of also to the engraver, who is not an Englishman, but one, that I find a very good workman, and very punctual to his word; which was the reason I did not employ Mr. Faithorne, as you directed, he having so very often and often disappointed my expectation.[72]

The public face of science as we know it begins to emerge here, in the last quarter of the seventeenth century (and with the energetic participation of Royal Society publicists like Hooke). An earlier tradition of images of individual inventiveness might have shown the isolated scholar, pen in hand, transferring his thoughts to paper.[73]

We don't notice the transfer of attention, from the mind to the measuring instrument that gave access to the raw data, which Hooke here takes for granted as the purpose of the illustrations for Boyle's book.

CHAPTER 3

SUBTLE ANATOMY

Very small animals

ONCE MASTERED, THE techniques of observation with a microscope gave even those without formal scientific training access to
remarkable discoveries. In spring 1673, Antoni van Leeuwenhoek, a
minor city official in the Dutch town of Delft, wrote what was to
be the first of hundreds of long, detailed letters (in Dutch) to the
Royal Society in London, describing meticulously, with pictures,
exactly what he was able to see through his home-made, single-lens
microscope. He made no pretence to expertise in anything apart
from the manipulation of his microscopes; he presented his observations as raw data, in day-by-day diary form, 'that they may be
better credited in England and elsewhere'.

  'I beg you and the Gentlemen under whose eyes this happens to
come to bear in mind that my observations and opinions are only
the result of my own impulse and curiosity,' he explained, with
studied humility. 'Take my simple pen, my boldness and my opinions for what they are; they follow without any particular order.'[1]
His fellow-countryman Constantijn Huygens senior (Christiaan's
distinguished diplomat father) sent along an accompanying letter of

introduction, further stressing that this material was offered unprocessed, for more eminent minds to interpret: 'He is a person unlearned both in sciences and languages, but of his own nature exceedingly industrious and curious. I trust that you will not be unpleased with the confirmation of so diligent a searcher as this man is, though always modestly submitting his experiences and conceits to the censure and correction of the learned.'[2]

We do not know how and when Leeuwenhoek began making and using single-lens microscopes. Like Pepys he was inspired to pursue this pastime of his intensively as a result of seeing the wonderful engravings of microscopic structures in Hooke's *Micrographia*; unlike Pepys, though, Leeuwenhoek had less easy access

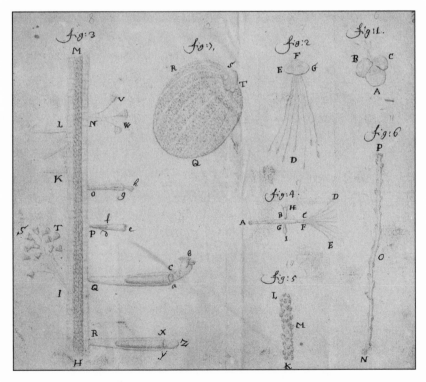

1. Original red chalk drawing of duckweed by Leeuwenhoek

to the accompanying English-language commentary, since he spoke no other language than Low Dutch. (Leeuwenhoek's letters were translated for the Royal Society by Oldenburg – a native of Bremen in Germany, who spoke French and Dutch – and Hooke, who taught himself Dutch for the purpose.)[3]

From his very first letter to the Royal Society Leeuwenhoek's interventions were observationally authoritative. He used the greater magnification of his 'recently invented microscope' [*mijn nieuw gevonden microscopix*] to make corrections to the representations of a bee's sting and a louse's body in Hooke's *Micrographia*; he detected 'blossom-like leaves' on the stalks of household mould. For fifty years, from 1673 until his death in the 1720s in his eighties, Leeuwenhoek assiduously observed and recorded the minutiae of the natural world, selecting his material for examination under the guidance and in accordance with the interests of the London natural scientists, and conscientiously forwarding his results to the Royal Society for scrutiny.

In his third letter to the Royal Society, Leeuwenhoek described millions of tiny, frantically squirming animals (*diertgens* in Dutch), which he had discovered in an infusion of pepper-water:

> I put about one third of an ounce of whole pepper in water, placing it in my study, with the sole design thereby that the pepper being rendered soft I might be enabled the better to observe what I proposed to myself. This pepper having lain three weeks in the water, to which I had twice added some snow water [the purest water available] (the other water being exhaled) I looked upon it the 24th of April, 1676, and discerned to my great wonder, an incredible number of very small animals of divers kinds.[4]

In London, the announcement of Leeuwenhoek's discovery at a Royal Society meeting in autumn 1676 caused quite a sensation, swiftly followed by disappointment as it was found that the Society's own experimenters (notably Hooke) could not replicate Leeuwenhoek's findings. Oldenburg wrote asking for further information on his 'method of observing'. Leeuwenhoek was characteristically guarded in the amount of detail he was prepared to go into in describing the equipment and techniques he used (we still do not know quite how he achieved the extraordinary magnifications shown in his drawings). It was not until a full year later, on 15 November 1677, that the indefatigable Hooke, having finally perfected Leeuwenhoek's technique of drawing off minute amounts of liquid with the thinnest of glass capillary tubes, successfully demonstrated the existence of the 'little eels' to the members at their weekly meeting:

> The first experiment there exhibited was the pepper-water, which had been made with rain-water and a small quantity of common black pepper put whole into it about nine or ten days before. In this Mr Hooke had all the week discovered great numbers of exceedingly small animals swimming to and fro. They appeared of the bigness of a mite through a glass that magnified about an hundred thousand times in bulk [about 450 times]. They were observed to have all manner of motions to and fro in the water; and by all, who saw them, they were verily believed to be animals; and that there could be no fallacy in the appearance. They were seen by Sir Christopher Wren, Sir Jonas Moore, Dr Grew, Mr Aubrey, and divers others; so that there was no longer any doubt of Mr Leeuwenhoek's discovery.[5]

Hooke himself notified Leeuwenhoek that the Dutch microscopist's observations were now considered properly scientifically verified, and informed him that no less a person than His Majesty the King had requested to be shown the 'animalcules', and 'was very pleased with the observation and mentioned your name at the same time'.[6] Leeuwenhoek became an overnight scientific celebrity, and was made an overseas member of the Royal Society in 1680.

Distinguished visitors, including several crowned heads of Europe, flocked to Delft to see the great 'searcher' in action: 'everyone is rushing to pay homage to Leeuwenhoek as the great man of the century', Constantijn Huygens the younger wrote to his brother Christiaan.[7]

What was Leeuwenhoek doing looking at pepper-water in the first place? As a typical amateur enthusiast, any recondite or unusual substance excited his curiosity. He examined his own urine and semen under the microscope, with fascination. The imported groceries on his dinner-table, which pleasurably excited the palate, were a predictable experimental choice. Besides, the spices being shipped in increasing quantities from Asia by the Dutch East India Company were a major contribution to the commercial wealth of the Low Countries. Anyone who could understand the origins of pepper's peculiar pungency to the tongue might hold the trade secret for manufacturing a local, less pricy alternative for European tables, and make his fortune.

The protozoa Leeuwenhoek found in his pepper-water were a by-product of these investigations of everyday materials. He was not actually looking for little animals when he made up his mixture of spice and water. Rather, he was trying to 'discover the cause of the pungency of pepper upon our tongue' by looking at the minute parts of the spice in solution. Like other self-styled 'mechanists', he believed that qualities of bodies, like taste, colour and smell, were a consequence of the shapes of their minute parts, and it was in pursuit of these distinctively shaped 'corpuscles' – which in the case of pepper he expected to be spiky and abrasive to the tongue – that he put pepper-water under his microscope.

Myriad, hitherto unimagined, wriggling animals were, however, more immediately arresting than any curiously shaped pepper-parts that the microscope might have detected (in any case, Leeuwenhoek could find no such prickly corpuscles). Not for the first time, small animals proved the most entertaining of specimens when viewed

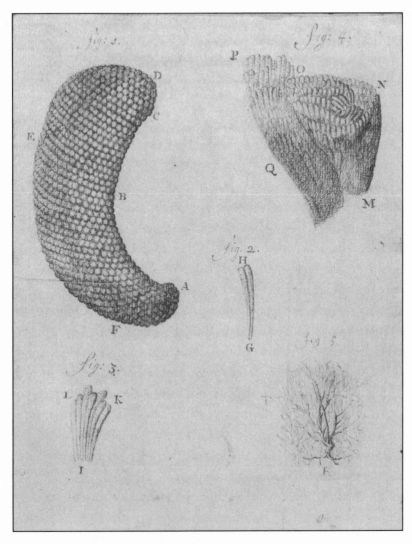

2. Original Leeuwenhoek drawing of a bee's organs

with a microscope. When Galileo had adapted his telescope to magnify tiny objects instead of distant stars, he began by observing houseflies; the earliest surviving microscopic illustrations, published in 1625 by Galileo's fellow academicians of the Accademia dei Lincei in Rome, showed the honeybee and its parts. Wren had first delighted Charles II with microscopically enlarged images of fleas and lice (and Hooke gave them pride of place in his *Micrographia*); when Christiaan Huygens began constructing his own instruments, he too chose to examine the minute structure of insects. According to honorary Dutchman René Descartes, all Europe was fascinated by 'flea-glasses' – small single lenses through which to observe the animal life crawling around the underwashed early modern man or woman's person.[8]

Descartes' views were important to the impressionable Leeuwenhoek – his mechanistic view of matter was based squarely on that of the Frenchman. Descartes had settled in the Low Countries in 1629, to pursue his philosophical thoughts uninterrupted by the hurly-burly of urban life in his native France. The more free-thinking Dutch intellectual atmosphere appealed to him; he had been deeply shocked by the Italian Catholic Church's condemnation of Galileo's astronomical work. Descartes' presence greatly influenced contemporary Dutch experimentalists, and scientific attitudes more generally, particularly his commitment to an entirely mechanistic explanation of nature – Descartes argued that all natural bodies are machines, whose smallest parts were as yet invisible to the inquiring eye. He held out hope for the discoveries of minute structure to be made with the microscope, and believed that the remarkable magnifications Leeuwenhoek was achieving with his glass bead microscopes promised access to the minutest of mechanistic parts. In London, meanwhile, Robert Boyle's comparable belief that improved microscopes could reveal the 'asperities' in the surfaces of bodies, which accounted for their colours, similarly encouraged instrument-makers towards combinations of lenses with ever-increased magnification (goaded on as always by Hooke).[9]

In The Hague, Christiaan Huygens also reacted with excitement to Leeuwenhoek's 'very small animals', of which his efficient information service (in the form of regular correspondents Moray and Oldenburg) quickly notified him. He began experimenting with his

3. Leeuwenhoek microscope

own bead microscopes to see if he, too, could replicate Leeuwenhoek's observations (helped by another Dutch microscopist, Nicolaas Hartsoeker).[10] By the middle of July 1678 Christiaan was back in Paris, where he had been appointed a permanent, salaried member of the Académie Royale des Sciences. There he revealed to the members the microscopic life Leeuwenhoek had found in pepper-water, and then published the first public announcement of the discovery in the Academy's journal, the *Journal des sçavans* – Huygens typically tended to rush into print to establish the priority of his own scientific activities over those of others. Early in August, John Locke, travelling in France, reported to

Robert Boyle 'the extraordinary goodness of a microscope Mr Huygens has brought with him from Holland'.[11]

Leeuwenhoek was an exceptionally skilled microscopist, and deserved the fame brought to him by his untiring observations with equipment guaranteed to ruin the eyes. Seeing phenomena clearly using an early microscope was difficult enough; recording what had been seen created problems of a different order again. In *Micrographia*, Hooke had described the subtle, time-consuming process by which the skilled microscopist transferred to paper what he saw through

4. Poppyseeds from Hooke's *Micrographia*

his lens, to produce the seductive images that made his and Leeuwenhoek's published works best-sellers. Before Hooke began the drawing of much-magnified mould spores, or of the cells in a thin slice of cork, he made repeated observations, under differing illuminations. His composite drawing was then passed to the engraver, who generally followed his drawing 'pretty well':

> In divers of the delineated Subjects the Gravers have pretty well follow'd my directions and draughts; and in making of them, I indeavoured (as far as I was able) first to discover the true appearance, and next to make a plain representation of it. This I mention the rather, because of these kind of Objects there is much more difficulty to discover the true shape, than of those visible to the naked eye. And therefore I never began to make any draught before by many examinations in several lights, and in several positions to those lights, I had discover'd the true form.[12]

The precious images that Leeuwenhoek painstakingly interpolated into the written texts of his letters, and which were so greatly admired by the Royal Society, were produced yet more contrivedly than Hooke's. Leeuwenhoek did not draw (whereas Hooke had trained briefly in London as a portrait painter). Instead, he used the services of a series of skilled Delft artists and draughtsmen to produce the illustrations he mailed to London. These were the only close collaborators Leeuwenhoek was prepared to tolerate (he was notoriously secretive about his microscope techniques); the unnamed Delft draughtsmen participated in Leeuwenhoek's exploratory observations, intervened to specify and clarify the detail as they recorded on paper what they mutually agreed had been seen, and were a critical voice at Leeuwenhoek's shoulder as he made sense of his microscopic images.[13]

Given the importance of visual material to activities in the new sciences, the Royal Society was particularly fortunate in having more

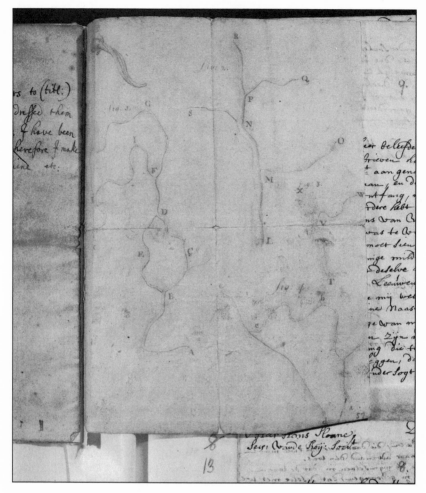

5. Leeuwenhoek drawing of capillaries in the tail of a tadpole

than one trained artist among its London members. England's most celebrated draughtsman, Christopher Wren, lent his steady hand to the precise recording of perishable natural phenomena as seen under the microscope and dissecting knife on several occasions. In the preface to his *Anatomy of the Brain* (1664), Thomas Willis paid tribute to Wren's participation, with medical doctors Lower and

Millington, in the dissections themselves, on which the book was based. They 'were wont frequently to be present at our Dissections, and to confer and reason about the uses of the Parts'. He went on:

> Dr Millington, to whom I from day to day proposed privately my Conjectures and Observations, often confirmed me by his Suffrage, being uncertain in my mind, and not trusting my own opinion. But the other most renowned Man, Dr. Wren, was pleased out of his singular humanity, wherewith he abounds, to delineate with his own most skilful hands many Figures of the Brain and Skull, whereby the work might be more exact.[14]

Wren did the drawings for a number of Willis's engravings of the dissected brains of men and animals, just as he had collaborated with Hooke on the drawings for *Micrographia*.

Microscopy was a tool that sharpened image-making in art and science. For both the Willis and the Hooke volumes, Wren used magnification with lenses for the fine detail of his line-drawings – in a letter to William Petty, he specified that drawings like these were 'more exactly trac'd by the help of Glasses [lenses]'. The techniques of northern European still-life drawing (for which lenses had long been in use) transferred easily to the new sciences, and brought another craft-skill into play. Fine craft work had long made use of magnifying lenses, and convex-lensed spectacles were widely available to correct myopia (Constantijn Huygens senior remarks in an autobiographical fragment that he used the spectacles he wore for short-sightedness to examine and magnify small insects).

In the 1670s, across Europe, artists, artisans and scientists came together in partnership to produce lasting records of ever tinier aspects of the natural world. Suspended as fine line-drawings on the printed page, without the clutter of scale-setting surroundings, the tiny living organisms depicted in seventeenth-century scientific engravings appeared arrestingly strange and beautiful. When Constantijn Huygens senior first looked at an insect through a lens

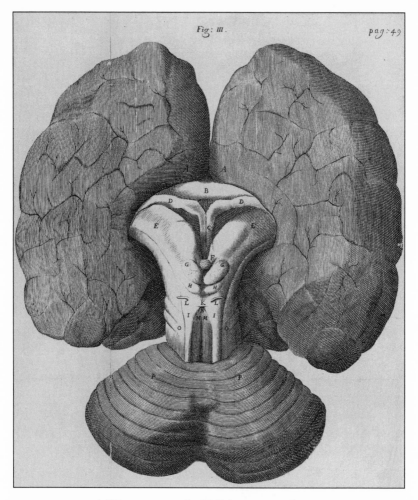

6. Wren engraving for Willis's *Anatomy of the Brain*

his immediate thought was to seek out the best contemporary artists to record the exquisite beauty of what he saw – the kind of Dutch artists who themselves already made use of magnifying lenses in their finely detailed still-life paintings of exotic flowers with *trompe-l'oeil* renderings of insects nestling among their petals.[15]

7. Dutch seventeenth-century flower painting with insects

Moving parts

The microscope lens revealed unimagined natural beauty to the eye; it also fuelled scientific speculation about the intricacy of internal anatomies on a tiny scale. It jarred the anatomist's imagination with images of the actual moving intestines and pulsating heart within the transparent flea or louse. Microscopists inevitably became dissectionists: Hooke graduated from thinly slicing cork and vegetable matter to slicing open the abdomens of flies in search of the vessels found in larger animals. When he discovered them in abundance he concluded that there was no less 'curious contrivance' here than in the larger animals. Henry Power also cut open a variety of insects, in some cases by simply cutting off the head and observing the beating heart even with the naked eye.[16]

The crucial novelty here for the anatomist was that the living insect under the microscope moved. The heart pumped, the digestive tract propelled its contents round the intestines under the gaze of the observing scientist. That animation could not, of course, be captured on paper even by the most talented of draughtsmen; it was, however, vital for an understanding of the mechanics of bodies.

Seventeenth-century scientists' fascination with the mobility of functioning, miniaturised internal organs, or 'subtle anatomy', matched their enthusiasm for another piece of seventeenth-century lens technology, the *camera obscura*. The simplest *camera obscura* is a darkened room, which light enters through a single small hole in (say) a window-blind. If a screen, or a piece of paper is placed so as to catch the entering rays, the viewer sees a full colour, moving image (upside-down) of the scene outside the window. Johannes Kepler improved on this amusement by incorporating a convex lens to enhance the image quality. In 1620 the diplomat Henry Wotton, travelling on government business in Germany, described Kepler's *camera obscura* in a letter to Francis Bacon:

> He hath a little black tent set up, exactly close and dark, save at one hole, about an inch and a half in the diameter, to which he applies a long perspective trunk, with a convex lens fitted to the said hole, through which the visible radiations of all the objects without are intromitted, falling upon a paper, which is accommodated to receive them.[17]

In his influential handbook on painting, the Dutch artist Samuel van Hoogstraten particularly stressed the 'cinematic' quality of *camera obscura* images:

8. A *camera obscura* in action

Before we leave the subject of reflections, I must say something about the picture-making invention with which one can paint by means of reflections in a closed and darkened room everything which is outside. I have seen this very thing done quite marvellously in Vienna at the residence of the Jesuits, in London by the river, and in other places as well. In Vienna I saw countless people walking and turning about on a piece of paper in a small room; and in London I saw hundreds of little barges with passengers and the whole river, landscape and sky on a wall, and everything that was capable of motion was moving.

The *camera obscura* provided vital 'illumination' for the young painter, he added, with studied play upon words. His aspiration should be to match the impact of such images:

9. Kepler's drawing of his improved *camera obscura*

I am certain that the sight of these reflections in the darkness can be very illuminating to the young painter's vision; for besides acquiring knowledge of nature, one also sees here the overall aspect which a truly natural painting should have.[18]

To capture on a screen precise images of distant figures, in full colour and motion, was a stimulus and challenge to the human imagination. It was also, inevitably, a vain pursuit — the gap was unbridgeable between the ephemeral animated screen-show and a static landscape with figures permanently recorded as a drawing or painting.

Illusionistic art, which depended on a scientific understanding of light, vision, perspective and reflection, like van Hoogstraten's own, was all the rage with the scientific community in London. Van Hoogstraten lived and worked in London from 1662 to 1667, and recounted with pride how he dined as an equal with several gentlemen members of the Royal Society at the house of Thomas Povey.

Povey owned two of van Hoogstraten's large perspective paintings. One of them, *Perspective from a Threshold* (1662), is a household interior, in which a long, black and white tiled corridor, complete with glimpses of furnishings and household pets, stretches, in exaggerated perspective, away from the viewer. Pepys records in his diary that Povey had *Perspective from a Threshold* hung in such a way that he could reveal it to guests by opening the door of his ground-floor closet. He describes how delighted and amazed he was to discover that no real corridor lay before him, and that 'there is nothing but only a plain picture hung upon the wall'.[19]

In the National Gallery in London there survives a further virtuoso piece of scientific art by van Hoogstraten, a 'perspective box', with viewing peep-holes, like an inside-out *camera obscura*. On the inside walls, floor and ceiling of the box are illusionistically painted scenes of the interior of van Hoogstraten's own family home. The peep-holes force the viewer to look with monocular vision, at an angle, at the scene inside. As a result, the eye constructs the scene into three-dimensional realism, while another illusionistic corridor apparently stretches deep into the field of vision, away from the viewer's gaze. A final curious consequence of this enforced peep-hole viewing is that the scene appears much bigger than it really is – just as microscopists regularly remarked that a sense of scale was lost altogether, peering through the lens at the object beneath.

In Paris, and London, Leeuwenhoek's newly discovered little animals joined a growing repertoire of subjects for 'subtle anatomy' – scrutiny and dissection under a magnifying lens. Tiny animals like Hooke's gnat and Leeuwenhoek's water-life were particularly attractive, doused in alcohol, or impaled on the end of a pin, under a miniature microscope lens, because of the effortless, mobile visibility of their inner moving parts. Hooke commented in *Micrographia* on how, when observing the water-gnat's organs, he could 'perceive, through the transparent shell, while the Animal surviv'd, several motions in the head, thorax, and belly, very distinctly'. Microscopes, he comments,

10. *Perspective from a Threshold* by van Hoogstraten, which Pepys saw at
Thomas Povey's house

allow non-interventive observation of internal organs, which could only be exposed using a dissecting knife on a larger animal:

> We have the opportunity of observing Nature through these delicate and pellucid teguments of the bodies of Insects acting according to her usual course and way, undisturbed, whereas, when we endeavour to pry into her secrets by breaking open the

11. Van Hoogstraten's perspective box

12. Vermeer's *View of Delft* (probably made using a *camera obscura*)

doors upon her, and dissecting and mangling creatures whil'st there is life yet within them, we find her indeed at work, but put into such disorder by the violence offer'd, as it may easily be imagin'd how differing a thing we should find, if we could, as we can with a Microscope, in these smaller creatures, quietly peep in at the windows, without frighting her out of her usual byas.[20]

Autopsy – seeing with your own eyes

The English physician William Harvey was one of those who observed, with delight, the steadily pulsating hearts of insects under a microscope. As he recorded in his landmark work *On the Motion of the Heart and Blood in Animals* (published originally in Latin as the *De motu cordis*, in 1628):

> In bees, flies, hornets, and the like, we can perceive something
> pulsating with the help of a lens; in *pediculi* ['little lice'], also,
> the same thing may be seen, and as the body is transparent, the
> passage of the food through the intestines, like a black spot or
> stain, may be perceived by the aid of the same magnifying
> lens.[21]

For Harvey, however, the value of such observations was as additional 'ocular' confirmation that the heart acts as a pneumatic pump, driving the blood in a perpetual circuit around the body of an animal, whatever its scale. The anatomical investigations on which he based that revolutionary discovery were conducted on much larger, red-blooded anatomies. As Physician Extraordinary, first to James I and then to his son Charles I, Harvey had access to a wide and exotic range of animals on whose living and dead bodies he could experiment. By 1616, when he gave the first indications of the theory of blood circulation in his Lumleian Lecture to the Royal College of Physicians, he had dissected not only human cadavers, but innumerable royal deer, and an ostrich from the King's menagerie; he had also conducted simple vivisectional experiments on domestic animals and excised the hearts of live vipers.[22]

Francis Bacon, Lord Chancellor to James I, educationalist and 'father of modern science', was at one time a patient of Harvey's. In his utopian fiction, the *New Atlantis*, he probably had the Royal Physician's anatomical dissections in mind when he described how his ideal scientists in the imaginary land of Bensalem kept 'inclosures of all sorts of beasts and birds, which they use not only for view or rareness, but likewise for dissections and trials; that thereby they may take light what may be wrought upon the body of man'.[23]

'I profess both to learn and to teach anatomy, not from books but

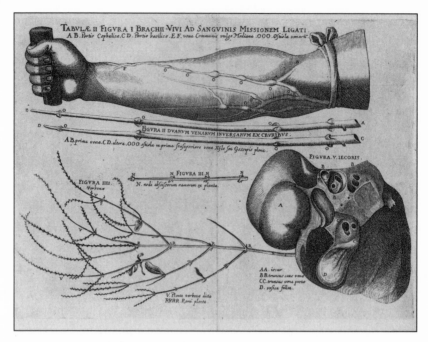

13. Harvey's illustration of an experiment with human veins, copied
from the work of his teacher Fabricius

from dissections,' Harvey announced boldly, in the letter to the
President of the Royal College of Physicians, with which he pref-
aced his *On the Motion of the Heart and Blood in Animals*.[24] Harvey set
himself the task of investigating 'the movement, pulse, action, use
and usefulnesses of the heart and arteries', an enterprise crucially
concerned with just that intricately interrelated movement of organs
observed less drastically in insects under the lens.[25] Like his Paduan
anatomy teacher Fabricius, Harvey regarded vivisection (the dissection of live animals) as the obvious best route to anatomical
discovery: he could not, of course, dissect the bodies of living
human beings, but he could, and did, carry out enormous numbers
of dissections of living 'brutes' (as he called them). These 'autopsies'
(literally, 'seeing for oneself') had to be repeated over and over again
in order for the anatomist to understand precisely how the body's

moving parts worked – exposing the heart to watch its movement, the lungs to see how movements of heart and lungs were related, severing the thorax to investigate the flow of air in respiration:

> For I could neither rightly perceive when and where dilation and contraction occurred, by reason of the rapidity of the motion, which in many animals is accomplished in the twinkling of an eye, coming and going like a flash of lightning. At length, and by using greater and daily diligence, having frequent recourse to vivisections, employing a variety of animals for the purpose, and collating numerous observations, I discovered what I so much desired, both the motion and the use of the heart and arteries.[26]

'The auricle contracts, and in the course of its contraction, throws the blood into the ventricle, which being filled, the heart raises itself straightway, makes its fibres tense, contracts the ventricles, and performs a beat.'[27] Like the moving image on the wall of the *camera obscura*, the exquisitely regular motion of the beating heart could only actually be seen under the anatomist's hand; however finely executed, no engraving or watercolour could ever capture it, no study of such pictures could reveal the complex regularity of that movement.[28]

Virtuoso vivisectionists

For the London Royal Society in its early years, Harvey was local hero and scientific role-model. He was also a convenient bridging figure – an arch-royalist, loyal friend and personal physician to Charles I, who had made a fundamental scientific breakthrough of the kind the post-Revolution Royal Society stood for. Abraham Cowley (who contributed an 'Ode to the Royal Society' to Thomas Sprat's 1667 *History of the Royal Society* too), in an enthusiastic if tasteless 1663 posthumous eulogy of Harvey, fantasises that the great anatomist pursues the coy nymph Nature deep into the human body, where she hides in the ventricles of the heart. Undeterred Harvey grasps 'this slippery Proteus', and forces her at knife-point to yield up 'her mighty mysteries' – the circulation of the blood.[29]

If the heart was a pneumatic/hydraulic pump, pumping a fixed quantity of blood around the body, what was the function of respiration, and what part did air play as that continuous circulation of blood passed through the lungs? It was to answer these fundamental questions, and in direct response to Harvey's discovery, that Boyle and Hooke developed the pneumatical air-pump. The 'spring' (PV = constant) experiments were a side-issue; what mattered was what happened to breathing creatures when gradually deprived of air as the glass globe enclosing them was emptied of air.

On the occasion on which John Evelyn watched 'divers experiments in Mr. Boyle's Pneumatic Engine' conducted by Boyle and Hooke at the Royal Society, these were experiments with animal respiration: 'We put in a snake, but could not kill it, by exhausting the air, only made it extremely sick, but the chick died of convulsions out right, in a short space.'[30] Cold-blooded animals, Hooke indeed discovered, required less air to survive than did warm-blooded ones. But if air was so essential for life, the question was, What happened to it inside the body? The logical next step, after countless numbers of birds and small animals had been suffocated inside the air-pump, while the members of the Royal Society solemnly timed their deaths, was to dissect live animals to try to observe respiration in process.

Hooke and Boyle vigorously promoted Harvey's model of endlessly repeated dissection as the one sure route to understanding the mechanics of respiration. In his early microscopical investigations, Hooke too had tried his hand at 'subtle' dissection under the microscope. But in matters of anatomy he was by temperament a large-scale hydraulic engineer rather than a miniaturist. None of the

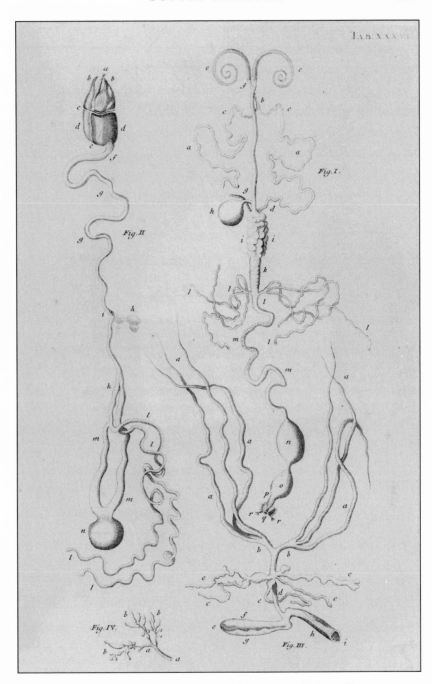

14. Swammerdam's illustration of the viscera of a butterfly

images presented in the *Micrographia* comes close to the dainty perfection of Jan Swammerdam's picture of the reproductive organs of a bee or the viscera of a butterfly. Hooke preferred to follow Harvey, and apply his dissecting knife to something more substantial.

Thus from air-pump suffocations, Hooke and Boyle graduated to vivisection, conducted on the respiratory organs of dogs (as Harvey had recommended in the *De motu cordis*). As usual, Hooke was the one who got his hands dirty; Boyle, more squeamish by temperament, preferred to hear about the autopsies at second hand. On 10 November 1664, in a letter to Boyle (who was away in the country), Hooke described an experiment in respiration performed on a live dog. He was pleased with the outcome of the experiment (which Harvey had himself described carrying out), less happy with the vivisection itself:

> The other Experiment (which I shall hardly, I confess, make
> again, because it was cruel) was with a dog, which, by means of
> a pair of bellows, wherewith I filled his lungs, and suffered
> them to empty again, I was able to preserve alive as long as I
> could desire, after I had wholly opened the thorax, and cut of
> all the ribs, and opened the belly. My design was to make some
> enquiries into the nature of respiration. But I shall hardly be
> induced to make any further trials of this kind, because of the
> torture of the creature; but certainly the inquiry would be very
> noble, if we could find a way so to stupefy the creature, as that
> it might not be sensible.[31]

This account contrasts strikingly in tone with the deadpan account of the same 'noble experiment' in the Royal Society's records: 'A Dog was dissected, and by means of a pair of bellows, and a certain Pipe thrust into the Wind-pipe of the Creature, the heart continued beating for a very long while after all the Thorax and Belly had been opened.' Evelyn, who also saw this experiment in the company of the Swedish Ambassador, agreed with Hooke that

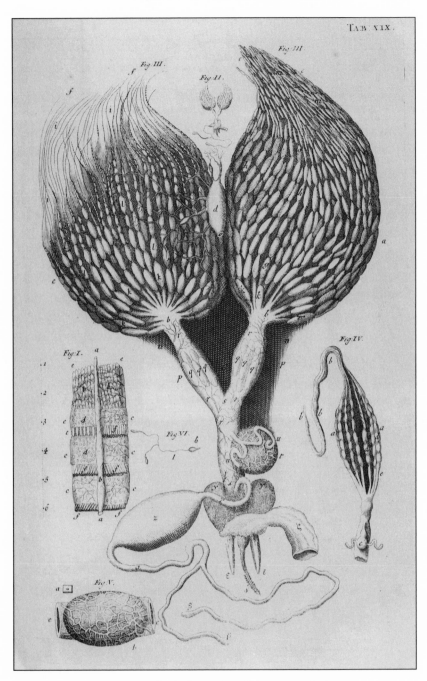

15. Swammerdam's illustration of the organs of a bee

'this was an experiment of more cruelty than pleased me'.[32] Hooke duly declined to repeat the experiment; however, the next two attempts at the dog and bellows demonstration at the Royal Society were thoroughly botched by less skilled dissectionists, and in the end Hooke once again took over the public performance of this experiment.

Harvey's demonstration that the blood circulated suggested all kinds of possibilities for experimentally exploring the further implications of that discovery. Among the members of the Royal Society, these focused on two ways in particular in which blood movement might be harnessed to promote bodily health: intra-venous injection and blood transfusion. Christopher Wren – as dextrous with the scalpel as with the pen – was acknowledged to be the most skilled English experimentalist conducting intravenous injection experiments.

Boyle described with gusto a pretty gruesome experiment in intra-venous injection that Wren carried out on a large dog at Boyle's Oxford laboratory:

> His Way (which is much better learn'd by Sight than Relation) was briefly this: First, to make a small and opportune Incision over that Part of the hind Leg, where the larger Vessels that carry the Blood, are most easy to be taken hold of: then to make a Ligature upon those Vessels, and to apply a certain small Plate of Brass (of above half an Inch long, and about a quarter of an Inch broad, whose Sides were bending inwards) almost of the Shape and Bigness of the Nail of a Man's Thumb, but somewhat longer. This Plate had four little Holes in the Sides, near the Corners, that by Threads pass'd through them, it might be well fasten'd to the Vessel; and in the same little Plate, there was also left an Aperture, or somewhat large Slit, parallel to the Sides of it, and almost as long as the Plate, that the Vein might be there exposed to the Lancet, and kept from starting aside. This Plate being well fastened on, he made

a Slit along the Vein, from the Ligature towards the Heart, great enough to put in at it the slender Pipe of a Syringe; by which I had proposed to have injected a warm Solution of *Opium* in Sack, that the Effect of our Experiment might be the more quick and manifest.

With equal cheerfulness Boyle went on to explain the way in which the opium injected by 'our dexterous Experimenter' was quickly carried 'by the circular Motion of the Blood' to the Brain:

So that we had scarce untied the Dog, (whose four Feet it had been requisite to fasten very strongly to the four Corners of the Table) before the *Opium* began to disclose its *Narcotick* Quality, and almost as soon as he was on his Feet, he began to nod with his Head, and faulter and reel in his Pace, and presently after appeared so stupified, that there were Wagers offered his Life could not be saved. But I, that was willing to reserve him for further Observation, caused him to be whipped up and down a neighbouring Garden, whereby being kept awake, and in Motion, after some Time he began to come to himself again; and being led home, and carefully tended, he not only recovered, but began to grow fat so manifestly, that 'twas admired.[33]

As always, the reports that come down to us describe successfully executed experiments of this kind. Where the dissectionist was less adept, things regularly went messily wrong. Pepys records going to the Royal Society with the Duke of York to see an experiment 'of killing a dog by letting opium into his hind leg': 'Mr Pierce the surgeon and Dr. Clerke [Timothy Clarke] did fail mightily in hitting the vein, and in effect did not do the business after many trials; but with the little they got in, the dog did presently fall asleep and so lay till we cut him up.'[34]

In July 1667 Dr John Wilkins brought to the attention of the

Society a French publication describing blood transfusions that had been carried out in Paris on two human subjects:

> The first was on a youth of fifteen or sixteen years of age, who had a violent fever and a lethargy upon him, his memory being lost, and little hope of life. After he had been blooded at a vein considerably, he had the arterial blood of a lamb transfused into him: and though before that he would be so sleepy, as scarce to be able to take sustenance, yet the next morning after this operation, he got up, and went about his business before five of the clock, and continued lively and well. The other experiment was on a labouring man, from whom ten ounces of blood having been taken, twice as much from the crural artery of a lamb had been put into him. The event was, that though after the experiment the man did not repose himself, as he had been desired, yet he was very well, and said, that he found himself more light and lively than he had ever done in his life before; and offered, that they should make that experiment upon him as often as they would.[35]

Physician Richard Lower had already been developing techniques for carrying out blood transfusions between dogs in Oxford. Not to be outdone by the French, the Royal Society immediately elected Lower a member, and proposed approaching Bedlam – London's lunatic asylum – for a suitable candidate for transfusion (the idea was that a change of blood would improve the patient's sanity). Lower, King and Hooke were duly dispatched to consult the physician in charge (he refused to co-operate). Meanwhile Oldenburg, corresponding with Boyle in Oxford, informed him that he had come up with the names of two possible candidates: one a gentleman who was 'a sad example of draining away too much of his blood (to allay the distemper of his brain), now emasculated, and for these many years changed from the most haughty and courageous temper to the most pusillanimous'; the other 'one Mr Thomas Hawker, who hath been outrageously distracted above a

16. Romanticised drawing of a sheep-to-man blood transfusion

year, upon whom two famed empirics have had their turns in trying the old ways in vain'.[36]

In the end it was another deranged gentleman, Arthur Coga ('a very freakish and extravagant man', according to Oldenburg, and an Oxford graduate in divinity), who agreed, for a fee of one guinea, to

allow himself to be used for a public demonstration of blood transfusion at a Royal Society meeting. The theory was that the blood
transfusion would cure his insanity (Coga worried later that it had
'changed his species'). The blood donor was to be a young sheep,
prompting Coga to quip that he was receiving 'the blood of the
lamb' and therefore, symbolically, the blood of Christ.

The experiment performed on Coga by Drs Lower and King on
23 November 1667, before a crowd of over forty witnesses assembled at Arundel House in the Strand, was, ostensibly, a success.
The anatomists let blood run for about a minute into a bowl
through a silver pipe fixed with a quill to the sheep's carotid artery
to calibrate the quantity of blood coming from the animal. King
tied Coga's arm, cut the skin over an adequately sized vein, opened
it with a lance and let out seven ounces of blood. He inserted
another silver pipe into the incision and joined the two pipes with
three or four quills. It took almost a minute before the lamb's blood
passed along the pipes 'and then it ran freely into the man's vein for
the space of 2 minutes at least'. King and Lower estimated that in
this time they had transfused about nine or ten ounces of blood into
Coga.' When at length Coga announced that 'he was not willing to
have any more blood', they stopped, pulled the pipe from his vein,
tied up his arm, and let more blood run from the sheep (to assure
themselves and onlookers that the flow had indeed been uninterrupted for the duration of the experiment).[37]

English scientific honour was satisfied. Coga reappeared before
the Society on 28 November, and 'produced a Latin paper of his
own, giving an account of what he had observed in himself since he
underwent the said experiment'. Secretary Oldenburg wrote to
Boyle, who was away from the Society at his own 'elaboratory' in
Oxford, with a detailed account of proceedings, and praising Dr
King, who 'performed the chief part of the transfusion with great
dexterity, and so much ease to the patient, that he made not the least
complaint, nor so much as a grimace, during the whole time of the

operation'. King himself reported (in a letter enclosed with Oldenburg's) that 'after the operation the patient was well and merry, and drank a glass or two of Canary, and took a pipe of tobacco in the presence of forty or more persons; then went home, and continued well all day, having three or four stools [bowel movements], as he used to have, his pulse being stronger and fuller than before, and he very sober and quiet, more than before'.[38]

The Coga transfusion experiment was deemed such a tremendous propaganda success for the Royal Society that it was repeated a fortnight later on 12 December, in front of an even larger crowd, including John Evelyn, who duly recorded the event in his diary.[39] This time, to achieve a more 'exact' trial, the doctors designed new pipes, weighed the unfortunate sheep before and after bleeding it, to determine the weight of blood transferred, and timed the transfusion. On this second occasion, according to King, things did not go quite as smoothly as on the first. He took what he guessed was about ten ounces of blood from Coga's arm, spilling some in the process. Because of the sheep's struggles, the quills kept slipping out of the receiver. After seven minutes, King estimated fourteen ounces had been transfused (much more than in November). Coga reported himself 'somewhat feverish' (Oldenburg maintained that he was drunk).[40]

Coga's drinking problem subsequently threatened to discredit these landmark transfusion experiments altogether. He was touted around the London coffee-houses as a scientific marvel, plied with drink, and encouraged to enlarge on his sad physical state and worse financial predicament since his notoriety. The Cambridge naturalist John Ray heard in January 1669 that 'the effects of the transfusion are not seen, the coffee-houses having endeavoured to debauch the fellow, and so consequently discredit the Royal Society and make the experiment ridiculous'.[41]

Coga may have been down on his luck, but in fact he was lucky to be alive. We now know that blood has to be carefully matched, even between humans, if transfusion is not to prove fatal.

At the end of 1667, the London scientists began to get wind of problems arising from the transfusion experiments the French had made so much of. Already on 8 October 1667 Oldenburg had reported to Boyle that 'the experiment of transfusion was tried at Paris upon a baron of Sweden; but he dying, his intestines were found all gangrened, so that it was not possible to have recovered him by any natural means. This invention is hugely disputed abroad pro and con.'[42] On 24 December, Oldenburg told Boyle that 'if these Parisians misrelate not, there hath been freshly made in that town an experiment of transfusion on a madman, with a surprising success' (did this perhaps mean that the November experiment at the Royal Society had, after all, not been the success it was first proclaimed?).[43] However, the following January, Oldenburg wrote that 'the madman, who found so notable a change by the transfusion, I think I told you of, to have been lately experimented at Paris, is somewhat relapsed, and pisseth blood; perhaps from too plentiful a transmission. It must be known both how much blood is transfused, and how much is fit to be transfused, to go sure.'[44] When the man eventually died, the Paris surgeons who had carried out the transfusions narrowly escaped a lawsuit.

The English dissectionists quietly abandoned their transfusion experiments on human beings as reports filtered in of the fatalities in France. Dr Lower, after his brief flurry of notoriety as the presiding surgeon at the London transfusions, went into lucrative private medical practice in Oxford, and dropped out of the activities of the Royal Society. He was expelled from membership for non-participation in November 1675, probably because he would not assist Hooke in repeating the dog and bellows respiration experiments, whose results had been contested by Dr Walter Needham.[45]

Political relations between France and England remained precarious throughout this period, while England was in open hostilities with the Netherlands. (During the days and nights when the Great

Fire raged in London, panic was increased by the belief that the fire had been deliberately set, either by the Dutch or by the French.) Oldenburg was himself briefly arrested and imprisoned in the autumn of 1667 on charges of passing intelligence to the enemy – a charge of which he was cleared, but which was obviously related to the vigorous correspondence he continued to conduct on behalf of the Royal Society with scientists from 'enemy' nations.[46]

Miniature marvels

Anglo-French scientific relations may have been strained at times – particularly around the issue of the 'transfusion-race' (precursor to the USA–Soviet space race). Foreigners of the Protestant faith, however, driven from their jobs in Catholic countries, were unequivocally welcomed by the English, and played a vigorous part in the new sciences. It was an Italian Protestant working in Basle who completed the medical breakthrough begun with William Harvey's discovery, by way of multiple dissections, of the circulation of the blood. Marcello Malpighi provided the missing piece from the jigsaw when he observed capillaries through his microscope, and identified them as the means whereby the blood pumped round the arteries by the heart was returned to the veins to begin the circulation process all over again.

Malpighi was a protégé of Oldenburg, whose work had come to the Royal Society's attention by way of the network of correspondence over which Oldenburg presided from the Society's foundation until his death in 1677. 'Our company of philosophers thinks that you are treading the real paths leading to a true knowledge of nature's secrets,' Oldenburg wrote to Malpighi. In strong contrast to the old, theorising philosophies, he continued, 'You devote your mind and hands to observing accurately and eviscerating minutely the things themselves.'[47]

In 1661, with the aid of his microscope, Malpighi observed the network of tiny vessels in the lungs, connecting the small veins to the small arteries, and thus completing the chain of circulation postulated by Harvey. Seven years later, the Royal Society sponsored publication of his treatise on the silkworm, in which, on the basis of microscopic examination of silkworm pupae from their cocoons, Malpighi observed vestigial wings, legs and antennae of the full-grown insect in the pupae, and argued that the eggs of all living things might contain the preformed materials for adult growth. The pupa, he wrote, 'is nothing but a mask or covering for the moth already engendered, so that if not harmed, it may acquire solidity and grow up like a foetus in utero'.[48] His research was made easier by the ready availability of silkworm pupae, as discarded material in the silk industry (like Harvey's 'little lice', the silkworm larvae have transparent bodies, allowing the anatomist to watch its internal organs in motion during vivisectional experiments).[49]

Waxworks

The speed with which organic materials deteriorated, in the days before refrigeration, hampered all dissection work. Soon after the middle of the seventeenth century, anatomists discovered that the fine vessels of specimens for dissection could be preserved (and made more visible) by injecting them with mercury, milk, ink, and a variety of other variously tinted liquids. The most successful, and dramatic, of these methods was one developed by the microscopist and anatomist Swammerdam. Swammerdam invented a way to inject warm, coloured wax, which solidified in the vessels, allowing the anatomist to examine them at his leisure. The technique was taken up and further developed by Frederik Ruysch in Amsterdam. Although Ruysch acknowledged that Swammerdam had been the first to have the idea, he insisted that his own technique was unique:

'What kind of substance this blessed man [Swammerdam] used to fill the vessels he never told me, and I never asked.'

Ruysch guarded the secret of his own procedure jealously. By 1700, he had developed wax injection to the point that he could claim that it appeared to restore the specimens to life, 'as all bear witness'. Visitors to Ruysch's display of preserved materials did indeed remark on the startling ability of the 'Ruyschian art' to restore and preserve the lifelike appearance of the human body and its parts, and Ruysch's museum of elaborate and often bizarre preparations was cited as one of the great wonders of the Dutch Republic.[50]

As usual, the new wax-injection techniques were brought to the attention of the Royal Society in London by Oldenburg, on the basis of correspondence received from Swammerdam in Leiden.

17. Detail of Ruysch title page showing his cabinet with specimens

On 12 June 1672, Oldenburg presented a sheet from Swammerdam's forthcoming work on the anatomy of the womb (*De uteri muliebris fabrica*) at a meeting of the Society, and read them the accompanying letter:

> This is to beg you earnestly to present this sheet on my behalf, and also be kind enough to give me an address for sending to you safely the human uterus and the other preparations of parts of the human body spoken of in this sheet.

The sheet in question contained the first two plates from Swammerdam's book, together with an account of his wax-injection process. In his letter, Swammerdam went on to explain that he proposed presenting to the Society the anatomical specimens on which these plates were based. He was, however, concerned at how safely to convey them, since the last thing he attempted to send to Oldenburg (a new book on the East Indies) had come back to him, because the ship 'being already at the mouth of the Thames, turned back because they heard rumours of war'.[51]

Oldenburg replied on 24 April. By now war had been officially declared between the two countries, and he agreed that it was a difficult time to be trying to transport scientific materials from the Dutch Republic to England. Undeterred, however, by the thought that he and his correspondent were now technically enemies, he pressed on with arrangements for the safe transport of Swammerdam's gift:

> We are disappointed because in these unfortunate times we cannot without difficulty obtain the uterus and other parts of the human body mentioned in your letter. The very learned Mr. De Graaf has also made ready several things for us, which are also held in suspense while there is doubt about a safe way of sending them. Yet I think that if those things, both yours and his, were entrusted to a certain Ostend merchant to be brought to London in some Flemish ship, they would come safely to our hands.[52]

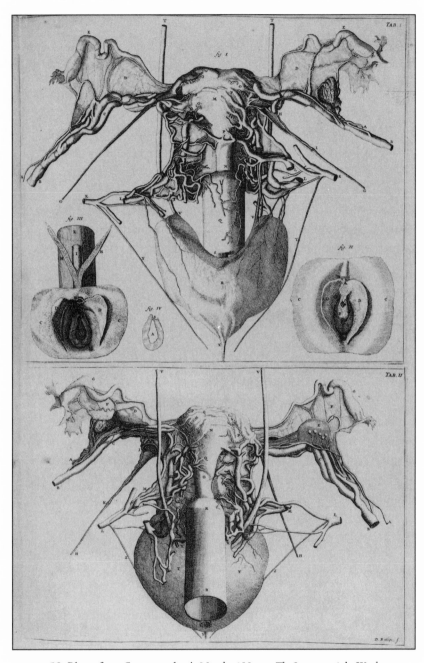

18. Plates from Swammerdam's *Miracle of Nature: The Structure of the Womb*

The wax-injected uterus finally arrived in December, and Oldenburg acknowledged it with gratitude, though he was obliged to notify Swammerdam that it had unfortunately been damaged in transit:

> After a chequered career and countless perils of war, your most elegant and skilfully prepared gift of a woman's uterus and other parts of the human body, with which you intended to enrich the Royal Society and its philosophical repository [museum], has finally arrived, famous Sir, at the beginning of this month of December. But it was delivered to me incomplete, for on one side of this box (that to which the uterus was attached), facing it, there was an empty space marked out by pinholes, to which I think it is clear that something was fixed, which has been removed. I found the rest sufficiently whole and undamaged, that is to say, the uterus of a woman in labour [puerperae], the pudenda of a virgin with the hymen, the veins and arteries of the human spleen, the branches of the hepatic artery, the arteries of a calf's spleen, the sphincter of the intestine of a fasting subject, the convoluted intestine of a ray, sections of the penis, urethra, clitoris, etc., together with the artery of the first kind in fishes (by which blood is brought to the gills), and also an artery of the second kind through which the blood [coming] from the gill vessels of the fish is directly distributed about the whole body, and lastly the peculiar lymphatic vessel from the abdomen of a cock – all things which are certainly very fascinating and prepared with exceeding ingenuity.[53]

The entry in the Royal Society Proceedings describes the 'human womb' as 'prepared after the method of Dr. Swammerdam, with all the other parts dried up, and the vessels filled with yellow and red wax, very distinctly injected, after the manner described by the doctor, in his book accompanying his present, which is dedicated to the Society'.[54]

Today we are bound to find Swammerdam's yellow and red wax

19. Display of bizarre wax-injected anatomy specimens by Ruysch

uterus less pleasing than Hooke's exquisitely engraved seeds and moulds, or Wren's meticulously observed lunar landscape. Nor do we consider Ruysch's bizarre arrangements to contribute anything convincing to the development of natural science. Swammerdam's specimens and Ruysch's displays seem to sit uneasily somewhere between freak-show items and research materials. They are, nevertheless, an intrinsic part of that motley fabric of which the seventeenth-century scientific activity was composed. Like the sheep-to-human blood transfusions, and the suffocations of small animals in the air-pump, they were crowd-drawers at meetings of the Royal Society. They encouraged members to look at the world around them with fresh eyes, employing, where necessary, the available new technology of the microscope. They contributed to the feeling that London was on the threshold of a new age, and commandingly at the centre of an international scientific enterprise of lasting importance – one with which neither political disputes nor even declarations of war could be allowed to interfere.

CHAPTER 4

RUNNING LIKE
CLOCKWORK

Against the clock

IF GALILEO REALLY did drop a feather and a cannon-ball from the top of the leaning tower of Pisa, he had no clock accurate enough to measure any difference in their relative times of descent. Precise measurement of tiny intervals of time has become so common-place to us today that it is hard to imagine a world without so much as a minute hand. Until well into the sixteenth century, clocks were set according to location for everyday purposes, reset by the sun at noon each day, and stopped and adjusted each time they ran fast and slow. Such were the discrepancies that accuracy to the near-est fifteen minutes was the realistic expectation. Towards the end of the sixteenth century, the brilliant Swiss clock-maker Jost Bürgi built William of Hesse a clock that could mark seconds as well as minutes — but it was a one-off, difficult to reproduce, and reliable seconds' measurement had to wait another hundred years.

Exact time measurement was the most absorbing technological challenge of the seventeenth century.[1] Without accurate time-keeping instruments, there could be no measurement of small increments of time elapsed between observations, and thus no consolidation of the advances in scientific understanding begun with the help of the telescope and the microscope. Increased precision in measurement of time elapsed was critical for three groups of people with practical involvement in applied science: astronomers, navigators and land surveyors. The clock was as indispensable as the telescope for producing reliable, tabulated data based on astronomical observation; it was an essential technical instrument alongside the compass and quadrant for determining position at sea, and alongside the level and theodolite for creating cadastral land surveys.

By the early seventeenth century astronomers were specifying star locations by two spherical co-ordinates: 'declination' (the star's angular elevation above a plane through the equator) as determined by one of the new sighting instruments, and 'right ascension' (the angle measured by the time elapsed between the star's passing through the meridian and that of the sun or other reference star). This method was first devised at Cassel by Christoph Rothmann, astronomer to William of Hesse – it was for him that Bürgi designed precision clocks to facilitate the new procedure. Initially these measurements were made using the naked eye; the rapid development of the telescope in the 1630s improved the accuracy of star measurements still further.[2]

In navigation, measurement of latitude and longitude were the established means of determining location away from recognisable coastlines and familiar landmarks. Ascertaining one's latitude was comparatively easy: north of the equator it was fixed by measuring the angle of the Pole Star above the horizon; navigators knew that one degree of altitude was equal to about fifty miles of distance north–south. In southern seas the sun's altitude could be used, with some adjustments, the calculations for which, by the sixteenth century, were available to navigators in printed tables.[3]

But knowing how far north and south you were was only part of the problem of fixing co-ordinates on the map. Determining one's east–west positioning – longitude – with confidence, became a

I. Seventeenth-century instruments used in mapping

matter of increasing urgency as navigators began, on a regular basis, to negotiate distant, dangerous waters like those along lucrative trade routes, westwards to Brazil and the West Indies, and eastwards around the Cape of Good Hope to the Spice Islands, India and China.

The principle of longitude measurement had been long understood; the problem was how to determine longitude reliably in
practice. Since the earth turns continually on its axis, there is nothing visible from one longitude that is not visible in the course of the
day from every other: along a given parallel the navigator observes
the same sun, the same moon, the same stars, but he sees them at
different times of the day. This means that the earth is a clock
(indeed, the original clock) and that longitude at any location can be
determined from time differences.

There are two obvious ways of calculating longitude: the time of
observation of a given celestial event (for example, an eclipse) at a
place of known longitude can be compared with the observed time
of the same event at the location whose longitude is sought; or the
navigator can keep constant track of the time at a place of known
longitude and compare that with local time. (In both cases local
time has to be computed by using the sun's position to set a clock to
local noon – not an altogether simple matter.)[4] In the long run the
use of an accurate chronometer to keep reference time (the exact
time back home, whose difference from the local time yields longitude at the desired location) proved more convenient and prevailed.
In the seventeenth and early eighteenth centuries, however, astronomical methods made the greatest headway, and promised the best
results.[5]

In the mid-1660s, Gian Domenico Cassini, Professor of
Astronomy at Bologna, observed the regular eclipses of the four
moons of the planet Jupiter (the so-called 'Medicean stars' discovered by Galileo), using an exceptionally fine telescope made by
instrument-maker Giuseppe Campani in Rome, and timed the transits with a pendulum clock. In 1668 Cassini published tables based
on his observations, predicting the precise moments of 'ingress'
(disappearance of a moon behind the disc of Jupiter) and 'egress'
(reappearance from Jupiter's shadow) of the moon's future eclipses.
These tables made it possible for two observers at separate locations

to observe and record times of ingress and egress simultaneously, and, by comparing their measurements against local time, to calculate the difference in longitude between them. In 1669 Cassini moved with his tables to a post at the Académie Royale des Sciences in Paris. There he presumably corrected his Bologna tables to Paris local time. A prime incentive for astronomers like Cassini accepting Louis XIV's invitation to join his Royal Observatory in Paris was the fact that in the 1660s the Paris meridian was one of the two most widely recognised international zero meridians, the other being London.

Between 1676 and 1681, a team of French astronomers from the Académie including Jean Picard and Philippe de la Hire, directly financed by the King, Louis XIV, carried out a complete land survey

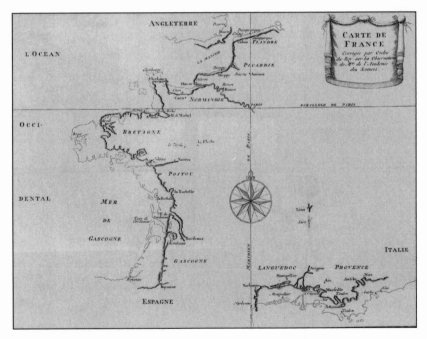

2. De la Hire's map of France superimposed on the outline of the old map

of France on the basis of Cassini's tables. The revised map of France was published by de la Hire in 1693, directly superimposed over a line-drawing of the best previous map of 1679, thereby vividly demonstrating the significant improvements in accuracy that the astronomical data had made. Because of the corrections produced by the survey, Louis XIV's kingdom was reduced in surface area by a fifth, prompting the King to remark, when he was shown the map in 1682, that he was repaid poorly by his astronomers for the support he had given them, and that he had lost more territory to them than to his enemies.[6]

By the 1680s the moons of Jupiter were the official French method of longitude calculation. In the same year that de la Hire published his map of France, Cassini reissued his tables, based now on almost three decades of observations. The accuracy of the motion of the first moon was now good enough to allow a conscientious observer to determine his longitude with an error of less than a degree (a distance at France's latitude of approximately 60 miles). Cassini added further refinements to his method, with the benefit of rapidly improving telescopes, providing increasing celestial detail for the observer:

> The eclipse of the first moon, which is faster than the others and which enters more directly into Jupiter's shadow, can be determined with the greatest precision; and after these eclipses of the moons one can use the shadows that they cast on the disk of Jupiter when they pass between the planet and the Sun, and also the permanent spots which often appear on the surface of Jupiter and which make a revolution around it which is the most rapid we have thus far discovered in the heavens, although the instant of the passage of these spots through the middle of Jupiter cannot be determined with the same precision as the instant of the eclipses of the moons.[7]

This process inevitably involves synergy − a two-way process in which the accurate clock helped both astronomers and surveyors to increase the precision of each other's work. Astronomers use the known latitude and the latitude of the location at which they observe the positions of celestial bodies to improve the precision of

3. Engraving of the French Royal Academy in session

these measurements. Surveyors and navigators use the agreed positions of the celestial bodies to improve the co-ordinates of locations
around the globe. By the mid-seventeenth century these two areas
were combined in the activities of individual astronomers and surveyors around observation of comets and eclipses, and territorial
surveying (particularly in the new colonies). Halley, for example,
collated observations of the 1680–81 comet at locations as distant
as the West Indies and the Cape of Good Hope. To do so he
needed the precise locations of the observation points, and the co-
ordinated times – in other words, knowing the longitude of the
location was essential.

On the other hand, those who had established land claims in
places like the new American colonies found that the unsophisti-
cated ways in which boundaries had been decided quickly led to
disputes between adjacent owners. Halley's interest in magnetic variation and in the fact that magnetic variation at any given point
varied over time arose out of disputes of this kind; territory originally surveyed using a simple compass method had to be resurveyed
because of local alterations in magnetic variation.

Following his own productive trip to St Helena, Halley had
repeatedly urged the Royal Society to take advantage of a transit of
Venus, which he knew would occur in 1761, to organise simultaneous measurements of the transit from points widely distanced on
the globe, and thence calculate the true distance of the sun from the
earth. Almost twenty years after Halley's death, two expeditions
were duly dispatched by the Royal Society, with funding from the
King, George II. The fifth Astronomer Royal Nevil Maskelyne
returned to St Helena to observe a yet more infrequent astronomical occurrence – a transit of Venus (the passage of the planet Venus
across the disc of the sun). Meanwhile two enterprising amateur
astronomers, Charles Mason and Jeremiah Dixon, set off for
Bencoolen in Sumatra. Only a day out of Portsmouth, however,
Mason and Dixon's ship was attacked by a thirty-four-gun French

frigate, leaving eleven dead, thirty-four wounded, and the ship considerably damaged.

Thoroughly rattled, Mason and Dixon announced that they were not prepared to go on, and suggested they might observe the transit of Venus from Iskenderun in Turkey instead. The Royal Society pronounced this mutiny, reminding the two men they were under contract, and had been paid in advance:

> Your refusal to proceed upon this voyage, after having so publicly and notoriously engaged in it, will be a reproach to the Nation in general; to the Royal Society in particular; and more especially and fatally to yourselves; and that, after the Crown has been graciously and generously pleased to encourage this undertaking by a grant of money towards carrying it on, and the Lords of the Admiralty to fit out a ship of war, on purpose to carry you on to Bencoolen; and after the expectation of this and various other nations has been raised to attend the event of your voyage; your declining it at this critical juncture, when it is too late to supply your Places, cannot fail to bring an indelible scandal upon your character, and probably end in your utter ruin.[8]

When the unfortunate Mason and Dixon reached the Cape of Good Hope they learned that the French had captured Bencoolen. Mason informed the Royal Society by letter that they would observe the transit of Venus at the Cape, and that they had landed their instruments to do so.

Meanwhile Maskelyne and his assistant Waddington had set up their instruments (including a fine pendulum clock by John Shelton, which Maskelyne noted ran consistently slow at its new location) two miles inland from Jamestown on St Helena, hoping that by choosing a lower observation point than Halley's they would avoid the troublesome, ever-present low cloud. At the precise moment at which they were to take their last crucial measurement, however, a cloud blew in and covered the sun, just as it had so often when Halley was observing almost a century earlier.

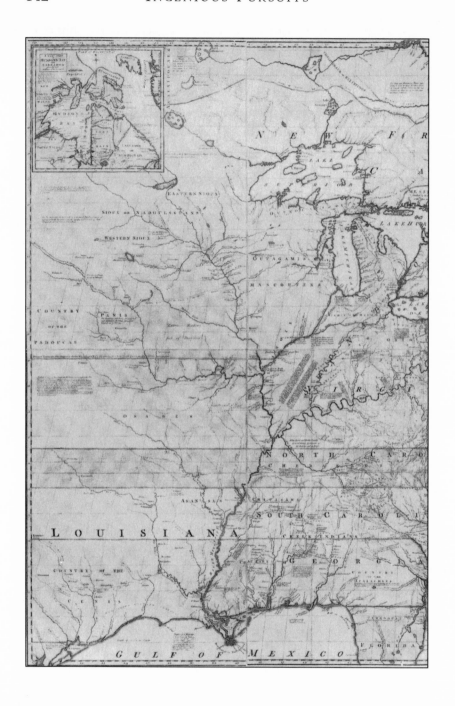

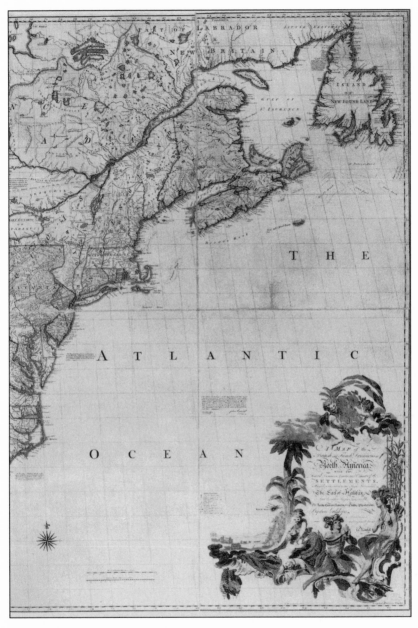

4. Mitchell's map of British and French territories in North America

Mason and Dixon stopped at St Helena on their way home, and reported that their observations at the Cape had been made with clear skies and near-perfect conditions. On their return to England they were promptly signed up again for their professional services (as an exceptionally successful travelling surveying team), and in November 1763 set off for North America to settle a long-standing boundary dispute between adjacent landowners Thomas Penn and Lord Baltimore, hereditary proprietors of Pennsylvania and Maryland respectively. The line of demarcation Mason and Dixon successfully established remains today as the Mason-Dixon line, which later became the boundary north of which African-Americans were free, and south of which they remained enslaved.

Having done what they had set out to do, Mason and Dixon wrote home to the Royal Society suggesting they use the opportunity, and their instruments, to establish the precise degree of latitude of Maryland, as a contribution to scientific investigation of the shape of the earth. The Royal Society duly sent additional instruments, including the Shelton clock that had already travelled with Maskelyne to St. Helena and on the Cape, and which subsequently travelled with Maskelyne to Barbados.[9] The most ambitious piece of mapping to come out of surveys like these was John Mitchell's 1755 map of British and French dominions in North America, updated many times in the course of the next twenty-five years.

Time trials

On 23 January 1675 (new style), Christiaan Huygens (also sponsored by the French Académie) drew a sketch in his notebook of a watch mechanism regulated by a fine coiled spring, and added, 'Eureka' – 'I have found it.' He was convinced he had found a clock-regulating device that was constant enough to ensure his clock kept good time over long periods, that was not affected by motion (unlike his earlier invention the pendulum), and that was small enough to make the resulting chronometer portable (suitable to carry on board ship). A week later he sent a sealed claim to priority for his important invention, in cipher, to Oldenburg at the Royal Society in London. Then, having discovered that the Paris watch-maker Isaac Thuret, whom he had employed to make up a working watch to his new design, was claiming the invention as his own on the basis of minor modifications he had made, Huygens changed his mind and decided to go public immediately.

So on 20 February 1675 Huygens followed up his first Royal Society communication with a letter to Oldenburg giving the solution to the cipher, and thus making his claim on the invention explicit: 'The arbor of the moving ring [the balance wheel] is fixed at the centre of an iron spiral.'[10] He went on to enlarge on his invention:

> The fact is, this invention consists of a spring coiled into a spiral, attached at the end of its middle [i.e. the interior end of the coil] to the arbor of a poised, circular balance which turns on its pivots; and at its other end to a piece that is fast to the watch-plate. Which spring, when the Ballance-wheel is once set a going, alternately shuts and opens its spires, and with the small help it hath from the watch-wheels, keeps up the motion of the Ballance-wheel, so as that, though it turn more or less, the times of its reciprocations are always equal to one another.[11]

Such a revolutionary clock could be made very small, making 'very accurate pocket watches'; larger models would be particularly useful 'for the discovery of longitude at sea or on land'. Huygens proposed that Oldenburg secure a patent for the invention either for himself or for the Royal Society (as a foreigner Huygens could not own a patent himself). At the same time, having secured a *privilège* for the watch in France, Huygens published an account of it (with diagrams) in the 25 February issue of the *Journal des sçavans* (the journal of the French Académie).

5. Christiaan Huygens

In London, Hooke learned of Huygens's claim to be the origi-
nator of the balance-spring watch a week later, while dining at
Boyle's house. He reacted swiftly and indignantly.[12] The very next
day he protested vigorously to the Royal Society (at a session at
which Flamsteed and Newton both happened to be present),

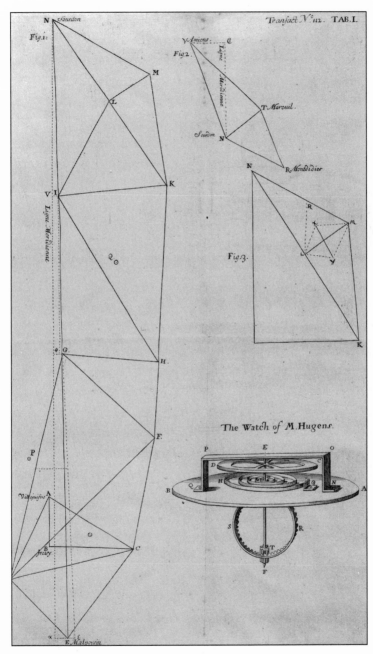

6. Drawing of Huygens's new watch, from the *Proceedings of the Royal Society*

reminding the members that he had presented them with his own invention of a spring-regulated watch in the early 1660s, and declaring that Huygens's spring was 'not worth a farthing'.[13] Nevertheless, to his consternation, the Royal Society, encouraged by Oldenburg, seemed inclined to accept Huygens's claim to priority.

Hooke had some justification for being annoyed. The Society's own archives documented the fact that he had produced a watch regulated by a small spring in the mid-1660s, in direct response to the challenge of devising a time-measuring device that could keep precise time on board ship. Indeed, the occasion for his proposal had been connected with an earlier horological invention by Huygens himself. For this was already the second time that the extraordinarily inventive Huygens had proposed an innovation in clock-design that promised to revolutionise time-keeping. In 1656 he had been the first person successfully to incorporate a steadily swinging pendulum as the isochronic oscillator in a clock.

In 1661 Huygens had visited London for the first time, and his new pendulum clock had attracted the attention of several members of the Royal Society with naval connections – particularly Sir Robert Moray, and a Scottish nobleman, Alexander Bruce. Bruce quickly spotted the possibilities of Huygens's pendulum clock as the precision instrument for measuring longitude at sea, and sponsored a series of tests to try out the effects of uneven motion on various designs of the clock, during sea-voyages, between London and The Hague (which Bruce carried out himself in 1661–2), London and Lisbon (conducted by Captain Robert Holmes in 1663), and London and Gambia (Holmes, 1663–4).

The early trials were not altogether a success. On the return leg of his trip to The Hague, Bruce was violently seasick, and failed to take proper care in keeping the clocks running; one of his clocks fell from its ball-and-socket style suspension and was damaged. However, enough progress was made with adjustments and modifications to successive models of the clock for Holmes to deliver favourable opinions of their usefulness; the trials also convinced the Duke of York (Admiral of the Fleet) and his cousin Prince Rupert of the Rhine (a senior naval commander) that the pendulum clock did indeed have a future as a marine chronometer – both were presented with pendulum clocks that they tested themselves.[14]

Meanwhile Huygens published – in Dutch – his *Instructions Concerning the Use of Pendulum-Watches for finding the Longitude at Sea* (1665), a practical handbook for seamen, which also included handy methods for computing local time at sea from timed observations of the sun.[15] The work was a canny piece of commercial marketing. Like Hooke specifying his microscopes in his *Micrographia*, Huygens included precise information about his latest pendulum clock as currently being tested, and the name of the instrument-maker in The Hague from whom it could be purchased.[16]

In March 1665 – within days of England and the Dutch Republic going to war (the second Dutch war), thereby making Huygens an enemy alien – a patent was granted to the Royal Society for 'marine timekeepers', specifically, the one developed by Huygens and Bruce. The deal was brokered by Sir Robert Moray; half of any proceeds were to go to the Royal Society, and the rest to be divided equally between Huygens and Bruce. A further series of sea-trials was carried out aboard a ship of the English Royal Navy. On 15 March Hooke reported back to the Royal Society (characteristically) that he was not at all impressed:

> Mr. Hooke remarked that in his opinion, no certainty could
> be had from pendulum watches for the longitudes, because,
> 1. They never hung perpendicular, and consequently the cheeks
> were false. 2. All kinds of motions upward and downward,
> (though it should be granted, that the watches hung in an exact
> perpendicular posture) would alter the vibrations of them. 3.
> any lateral motion would produce yet a greater alteration.

Viscount Brouncker, in his double role as President of the Royal

7. Frontispiece to Sprat's *History of the Royal Society* with Huygens's triangular pendulum watch among the inventions shown in the background

Society and Assistant Comptroller to the Navy Board, ordered that the watches under trial should be brought ashore, and the tests continued, as a matter of urgency, in rapidly moving carriages: 'some further experiments should be made with them, by contriving up and down motions, and lateral ones, to see, what alterations they would cause in them'.

Hooke (wildly competitive whenever an innovation in technology was at issue) now declared that he had designed a better alternative marine timekeeper and 'that he intended to put his secret concerning the longitude into the hands of the President, to be disposed of as his lordship should think fit'.[17] While the trials of the pendulum clocks had been going on, Hooke had come up with a proposal for some sort of compact clock or watch regulated by the isochronous movement of a spring rather than that of a pendulum. Brouncker had briefly registered Hooke's intention to the Royal Society in 1664: Hooke 'had discovered' to Brouncker, Moray and Bishop John Wilkins (a long-standing friend of Hooke) 'an invention which might prove very beneficial to England, and to the world', and the Society agreed to provide up to £10 to cover the costs of trials.[18]

The idea was that Lord Brouncker would negotiate lucrative terms for a patent on Hooke's alternative marine timekeeper under the same rubric as Huygens's. Given that a patent would guarantee a percentage payment to the Royal Society for every such timepiece manufactured and marketed, Huygens's pendulum clock and Hooke's spring-balance clock together promised to relieve the Royal Society's worsening financial situation (the Society's charter had made no provision for regular funding beyond members' donations). The vice-president, Sir Robert Moray, writing to Huygens himself the following year, acknowledged that surrendering the secret clock to the Royal Society had been an act of generosity on Hooke's part.

By summer 1665 Hooke was negotiating the detail of an

appropriate patent (the draft document survives at Trinity College, Cambridge). In addition to Wilkins, Hooke's other trusted friend, Boyle, was apparently taken into his confidence at this time.[19] But in the midst of these discussions, in late summer 1665, a devastating outbreak of plague in London forced the members of the Royal Society to disperse hurriedly to the country (economic activity in London was brought to a standstill for almost six months). Boyle and Moray retired to Oxford; Hooke and Wilkins went to Durdans, near Epsom, the seat of Lord Berkeley, taking a quantity of scientific instruments with them, including Boyle's air-pump, so that they could continue working.[20]

Shortly after their arrival, John Evelyn, a frequent visitor at Durdans, found them hard at work, 'contriving Chariotts, new riggs for shipps, a Wheele for one to run races in, & other mechanical inventions'. He wrote admiringly that nothing approaching the investigating power of this little group was to be found elsewhere in Europe, 'for parts and ingenuity'. They had set themselves an ambitious programme of experiments to carry out during their country recess: 'divers experiments of heat and cold, of gravity and levity, of condensation and rarefaction, of pressure, of pendulous motions and motions of descent; of sound, of respiration, of fire, and burning, of the rising of smoke, of the nature and constitution of the damp, both as to heat and cold, driness and moisture, density and rarity, and the like'.[21]

Away from the elaborate protocol and posturing of London Royal Society meetings, Hooke and Wilkins seem to have spent a particularly fruitful – and enjoyable – time practising science together during this period of enforced vacation. The new carriage they were working on (which Huygens sketched in his notebooks) was designed for speed and manoeuvrability. One can picture them careering around Durdans in a prototype coach, continuing the interrupted trials of the effects of violent motion on pendulum watches.

Hooke continued to work on watches with springs, as indeed he had been instructed to do by the Royal Society (it may have been while they were at Durdans that he gave Wilkins an early version of his spring-balance watch as a gift). Like Huygens during the same period, he also worked on methods and instruments for finding local time at sea (the crucial calculation needed alongside the time at home 'held' by the longitude timekeeper). And he devised a reflecting quadrant (of a kind eventually brought into regular use in the 1730s) for finding distances by triangulation (fixing places by intersecting rays). He and Wilkins tested this using Hooke's favourite experimental site, Old St Paul's. Hooke reported to Boyle in Oxford on 15 August 1665:

> My quadrant does to admiration for taking angles, so that thereby we are able from hence to tell the true distance between Paul's and any other church or steeple in the city, that is here visible, within the quantity of twelve foot, which is more than is possible to be done by the most accurate instrument, or the most exact way of measuring distances.[22]

Hooke was scheduled to present his 'new perspective for taking angles by reflection' to the Royal Society on 12 September 1666 – the week following the Great Fire of London. The meeting did not, of course, take place.[23]

The 1665 patent submission for Hooke's balance-spring watch was never lodged. Hooke said later that the patent negotiations foundered because of a final waiver sentence, which conceded that anyone who developed Hooke's idea further during the seven-year patent period was entitled to the financial proceeds of the improvement.[24] If the patent was to extend only to watches precisely like the prototype, it hardly seemed worth applying for.

Ten years later, in 1675, therefore, Huygens's claim that he was now entitled to a patent for himself, for a spring-driven watch remarkably like Hooke's own, seemed to Hooke sheer hypocrisy, as well as a personal insult. Hooke was convinced that the

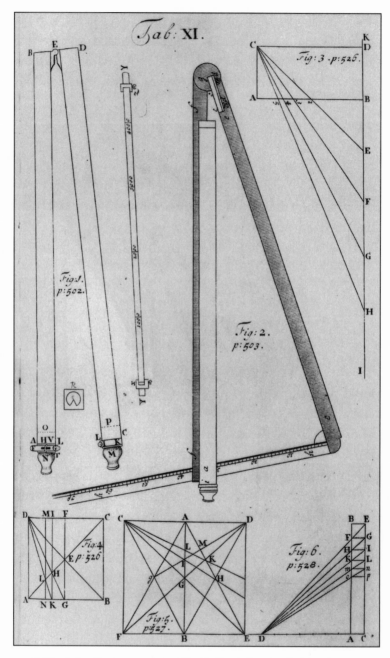

8. Hooke's reflecting quadrant (Fig 2); Fig I shows a double telescope
designed by Wren

breakthrough technical insight underlying such a watch was his own, and that Huygens had been made aware of it as early as 1665. And, indeed, surviving letters do show that Moray wrote to Huygens in Paris in September and October 1665 (not long after he had brokered his English patent for the sea-going pendulum clock) informing him of Hooke's balance-spring prototype as something that would be of particular interest to him in his efforts to perfect his marine chronometer. Moray acknowledged that he had been sworn to secrecy as part of the on-going patent discussions, but made the excuse that Hooke had chosen subsequently to lecture publicly on the spring watch at Gresham College, and that therefore the information was in the public domain. Oldenburg was party to this 'leak', since it was he who forwarded the letter (sent to him unsealed for approval), together with comments of his own, to Huygens.

In the course of 1675 the priority dispute between Huygens and Hooke developed into something between a melodrama and a farce – though Hooke himself treated the whole matter as an out-and-out conspiracy. Barely a year after England had called an ignominious halt to the third Anglo-Dutch war, with the country increasingly hostile to her Catholic neighbour France, the Royal Society chose officially to back a Dutch scientist working for the King of France, against their own Curator of Experiments in pursuit of a patent claim for a piece of important naval technology. Indignantly Hooke publicly accused the person he believed had 'leaked' details of his own balance-spring watch mechanism to Huygens of being an 'intelligencer' – a spy for the enemy.[25]

In spite of Hooke's objections, Oldenburg pressed ahead with Huygens's patent. As drawn up it would clearly have severely hampered attempts by any English competitor to produce timepieces working on similar principles:

Whereas we have been informed by the humble Petition of our Trusty and Wellbeloved Henry Oldenburg Esqr Secretary to the Royall Society that the Sieur Christian Hugens hath newly invented a certain Sort of Watches usefull to find the Longitudes both at Sea and Land, and hath transferred his interest therein upon him the said Henry Oldenburg for these Our Dominions; And having therefore humbly besought Us to grant him the Sole Right of making and disposing of such Watches within our Dominions. We being willing to give all encouragement to inventions which may be of publick use and benefit; Our will & pleasure is that you prepare a Bill for our Royall Signature to passe Our Great Seale containing Our Grant and Licence unto the said Henry Oldenburg and his Assignes of the soul use and benefit of his said Invention for the terme of fourteen yeares.[26]

Fortunately for Hooke, however, Sir Jonas Moore, chief surveyor at the Ordnance Office, stepped in at this point on his side of the dispute, and narrowly prevented the patent being granted to Huygens in April. During April and May, Moore advised Hooke closely on how to proceed, in order to get proper recognition for his own priority claim. Moore also lobbied at court on Hooke's behalf (he had tutored the Duke of York as a boy, and remained close both to him and to his brother, the King).

Since the day the Huygens patent affair erupted, Hooke had been hard at work with Thomas Tompion on an improved, working spring-balance watch of his own. On 7 April (old style) Moore arranged for Hooke and Tompion to present this prototype watch to the King himself; it was delivered in early May, inscribed some-what presumptuously 'Hooke inv. 1658. Tompion fecit 1675' (Invented by Hooke in 1658; made by Tompion in 1675). Hooke recorded in his diary: 'With the King and shewd him my new spring watch, Sir J. More and Tompion there. The King most graciously pleasd with it and commended it far beyond Zulichems [Huygens's].

He promised me a patent and commanded me to prosecute the degree. Sir J. More beggd for Tompion.'[27]

Over the following months the King personally 'tested' the Hooke/Tompion watch, reporting on its accuracy, returning it for modification (there seem to have been several models) and correction (when Charles reported the watch ran slow). In between times the King entered into the spirit of Hooke's conspiracy theory by keeping it 'locked up in his closet'. Oldenburg complained in a letter to Huygens in June that Hooke's watch 'is still hidden from us, only the King has seen it, and he refuses to make it public until the matter of the patent is settled'.[28] Meanwhile, in a symmetrical move, Huygens was being pressured by Oldenburg to send over a prototype of his watch for Brouncker to 'try'. It was a long time coming, and did not run well when it did finally arrive (in late June). Moore reported to Hooke that it 'wanted minutes and seconds' (had neither minute hand, nor second hand).[29]

Funded by the military

In fact, the King was not the only person testing the top-secret Hooke/Tompion watch. Sir Jonas Moore had one too. At Moore's request, Hooke had Tompion modify one of Moore's own pocket timepieces to incorporate a hair-spring regulated balance.

Moore's interest in clocks and watches was strictly tied to his military and naval responsibilities at the Ordnance Office. By training he was a mathematician and surveyor. In the 1650s he had played a leading role in the successful surveying and draining of the Fens, as surveyor-general to the fifth Earl of Bedford's Fen Drainage Company. He thus had extensive practical experience of the importance for precise determination of position on land of accurate ephemerides (tables of the positions of stars, eclipses, high and low tides etc., for a given year, from which local time, latitude and

longitude could be reckoned). In the early 1670s the young John
Flamsteed published annual English ephemerides. In 1674 Moore
collaborated with Flamsteed (with whom he had been put in contact by the antiquary Elias Ashmole) on a more ambitious
ephemerides, put together by Nicholas Stephenson (it appeared
under Moore's direction each subsequent year until Moore's
death).[30] The official title of this 'Royal Almanack' – as the
ephemerides was now officially called, since it was sponsored by the
King himself – vividly captures the contents:

> A Diary of the true Places of the Sun, Moon, and Planets,
> their Rising, Southing, and Setting, High-water at London-
> bridge, with Rules to serve other Places after the New Theory
> of Tides and Directions of Sir Jonas Moore . . . The Eclipses,
> Tables of the Sun's Rising, Moon's Southing, Moon's Rising
> and Setting [separately compiled by Flamsteed]. A Table of the
> Sun's Right Ascension in time for every day at Noon, and of
> thirty one of the most notable fixed stars. With the Moon and
> other the Planets Appulses to the fixed Stars . . . All done with
> great Pains, according to the Rules of Art, for his Majesty's
> Use; and at his Command.[31]

Accurate clocks had a vital part to play in the quality of the predictions in these ephemerides. When Moore used his considerable
collection of telescopes to make astronomical observations, Hooke
was on hand to take care of the clocks. He was consulted over the
setting of Moore's pendulum clock in connection with his tide
tables of 1673;[32] while on 1 January 1675, Moore, Edmond
Halley, Thomas Streete and Hooke used one of Hooke's balance-
spring watches when they observed an eclipse of the moon together
from the Tower of London.[33] They presumably checked the time of
occurrence of the eclipse against the data provided in the
Moore/Flamsteed almanack.

Between 1672 and March 1674 Moore, in his official capacity as

Surveyor-General of the Ordnance Office, had his hands full with supplying armaments to the English Navy during the third Anglo-Dutch War. Yet again the English fleet failed to distinguish itself in naval engagement after naval engagement. It began to be loudly suggested in opposition circles that the problem lay with the Ordnance Office: its officials, it was said, were in the pocket of James, Duke of York. James – who openly declared himself a Catholic in 1673, and subsequently resigned his post as Lord High Admiral – was said to have deliberately withheld essential military supplies from the English Navy under the command of Prince Rupert (James's cousin) during a series of naval engagements off the Schonveld.

The Ordnance Office was forced to issue an official denial of the story, including its own records as supporting evidence. A document issued on 27 May 1673 showed that everything Prince Rupert had ordered in the way of ammunition had been supplied to the letter. Moore himself had sailed out with six great ammunition ships filled with gunpowder and shot, and had met Prince Rupert at the Nore. Far from there having been a shortfall in supplies, a large quantity of gunpowder had remained after the battle was over, and had been returned to Portsmouth. The report cleared Moore's name, and incidentally put on the record that he was the last surveyor-general of the Ordnance Office to see active service.[34]

Moore's personal view was that the English Navy's poor performance should be put down to the poor standard of astronomical data currently available to sea-captains to enable them to navigate efficiently, and manoeuvre strategically. By 1674 Moore had decided that the way to produce the kind of reliable, accurate tables of data the English Navy needed, was to 'cause an Observatory to be built, furnished with proper instruments, and persons skilful in mathematics, especially astronomy, to be employed in it'.[35] He already had his eye on Flamsteed as the skilful astronomer for the job.

Immediately after hostilities with the Dutch Republic ended, in March 1674, Moore wrote to Flamsteed apologising for the neglect of his protégé, and giving 'the excuse of our much busines in the Warr'. He now offered Flamsteed ten pounds a year out of his own pocket, plus accommodation, and declared that he would 'have the best Pendulum Watch that can be made for yow' based on Huygens's design, plus all the astronomical instruments he needed, whatever the cost. In October he wrote to Flamsteed with a proposal (in which he indicated Hooke had had a hand) that Flamsteed should be set up in an observatory under the auspices of the Royal Society.³⁶

The following March, however, Moore gave up on the Royal Society (which he told Flamsteed he believed to be in decline) and resolved the matter himself with admirable efficiency. After an audience with the King it was agreed that Flamsteed should have an annual pension of £100, paid by the Ordnance Office. His brief would be to apply 'the most exact Care and Diligence to rectifying the Tables of the Motions of the Heavens, and the Places of the fixed Stars, so as to find out the so-much desired Longitude at Sea, for perfecting the art of Navigation'. The Ordnance Office was also to supervise the construction of an observatory at Greenwich at a cost of up to £500, to be responsible for maintenance and any other costs incurred, and to provide assistants. The £500 was raised, fittingly, by selling off 690 barrels of surplus gunpowder from the six ship-loads Moore had taken out to supply Prince Rupert during the third Dutch war two years earlier. Materials for the observatory itself were also old Ordnance stock: bricks from a surplus store at Tilbury, second-hand wood, iron and lead from demolition work at the Tower. In other words, the venture was entirely funded by the military.

Flamsteed moved to London from Derby immediately the deal was struck. While the observatory (designed jointly by Hooke and Wren) was being built, Moore found him quarters at the Tower of London, where Flamsteed had the use of a turret of the White Tower as his observatory. There, in May 1675, Moore, again under

9. Greenwich Observatory by Robert Thacker and Francis Place. Thacker was the Ordnance's sketcher of fortifications

terms of strictest secrecy, passed his own prototype of Hooke's top-secret balance-spring watch into the hands of Flamsteed to test, Flamsteed having observed that such a portable watch 'being set by a corrected Pendulum' would be 'of very good use in making observations'. Flamsteed used the watch for most of May, and found that it performed well (differing by only one minute per day from his pendulum clock).

Flamsteed was not, however, let into the secret of precisely how Hooke's watch worked. He wrote to his friend Richard Towneley, fellow astronomer and instrument-maker (inventor of the telescope micrometer-eyepiece), on 27 May 1675:

> You are desirous to know the Contrivances of Mr Hookes watch But I must confesse to you his humour is such hee would never suffer mee to see it tho hee made mee severall promises, onely I once had a view of the outside, of the inner workes: what Sr Jonas Moore tould me or I learnt by discourse with Tompion I enformed you formerly.[37]

Evidently Towneley too was anxious to enter the race to produce an acceptable working model of such a watch, since Flamsteed continued: 'I have hopes now you have got one patterne of [Huygens's watch] you will much amend it by your contrivance & that at length that will helpe us to one better than Mr Hookes.' Armed with the diagrams and text Huygens had published in the February issue of the *Journal des sçavans*, a skilled instrument-maker like Towneley could build and modify his own equivalent design as a candidate for a patent-worthy timekeeper.

Shortly afterwards Flamsteed explained to Towneley that the reason he could not examine the mechanism of the watch he was using (which he found to 'goe well to admiration') was because 'Mr Tompion had fild of the spring that closes the watch so that I could not open it to see how the spring was fastened within'. He also warned Towneley not to let Moore know they were discussing the watch since he and Hooke 'engage as much as they can to secrecy'.[38]

On 8 June Flamsteed was finally able to tell Towneley that the balance by which Hooke's watch was regulated was 'a payre of fine springs such as are used in the Childrens toyes to force a puppet out of the box when the lid which holds it downe is removed. These as far as I can understand lay hold upon the Axis of the ballance & theire vibrations with equall force being nearly aequitemporaneous keepe the motion equall. Sr Jonas told mee last night that the rimme of the Ballances so far as they play are toothed.'[39]

In the end, neither Huygens's nor Hooke's balance-spring watches proved accurate enough long-term to be serious contenders as marine timekeepers suitable for finding longitude by the time comparison method. Once this became apparent, Oldenburg's interest in Huygens's balance-spring watch petered out, and he (and Huygens) turned their attention once again to modified pendulum clocks as the most likely candidates. But Hooke's watch's limited accuracy over extended periods of time did not bother Sir Jonas Moore at the Ordnance Office. As far as Moore was concerned, his support for the balance-spring watch was based on its suitability (confirmed by Flamsteed's trials) as a portable timepiece, for use in the practical business of taking time measurements associated with astronomical observation — he was sufficiently impressed that Hooke's watch could be relied on to keep perfect time over a period of hours, even when hand-held and subjected to sharp movements.

Such a watch was the perfect accessory/auxiliary to a large, precision pendulum clock when telescopic observations were being taken of a lunar eclipse, the moment of disappearance of one of Jupiter's moons, or the transit of Venus across the face of the sun. Instead of constantly turning from the telescope to the pendulum clock (secured perfectly vertically against the observatory wall), the astronomer had only to deflect his glance slightly from his eyepiece, to establish and record the time of the successive moments of

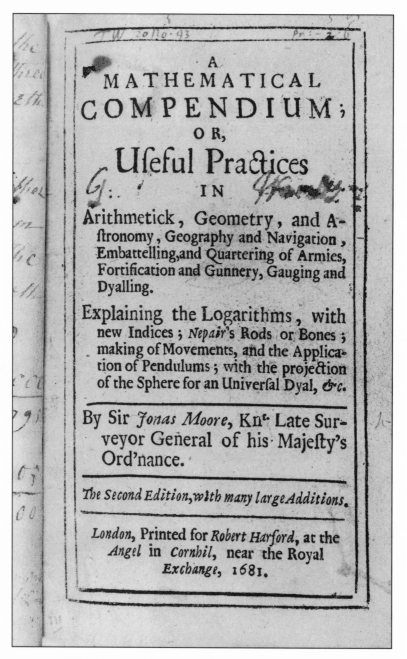

10. Title page of Jonas Moore's *A Mathematical Compendium*

the event under observation on his pocket-watch (set at the outset from the fixed clock).

These were the activities, patiently repeated for nights on end, needed in order to put together the star charts and ephemerides on which Moore pinned his hopes for the accurate calculation of longitude at sea. Such tables would (Moore believed) enable the mathematically trained navigator to calculate longitude on the basis of observation either of eclipses of the moons of Jupiter (particularly the first moon), or of the angular distance between the moon and a fixed star (the moon moves, like the hand of a clock, nightly across the sky). It was a view shared by some of the most prominent mathematicians and astronomers of his day. As late as 1725 Newton wrote confidently to the secretary of the Admiralty: 'The Longitude will scarce be found at sea without pursuing those methods by which it may be found at land. And those methods are hitherto only two: one by the motion of the Moon, the other by that of the innermost moon of Jupiter.'[40] A clock, Newton continued, can only *keep* time, it cannot *find* location. 'I have told you oftner than once', he wrote to another correspondent, 'that Longitude is not to be found by Clock-work alone. Nothing but Astronomy is sufficient for this purpose.'[41]

Nevertheless, Moore's official Ordnance backing for the two joint ventures undertaken together and proposed simultaneously to the King in March 1675, was, from his point of view, an unqualified success. He had presided publicly over the setting up of the Greenwich Observatory (with Flamsteed as Astronomer Royal), and a highly visible bid for recognition for a compact, portable timepiece (to the design of an Englishman). The Surveyor-General had shown technological discernment and knowledgeability, and had played a leading role in promoting the priority of English expertise in astronomy and instrument-making, at home and abroad. In the long run this more than compensated for any blot on his reputation caused by rumours concerning the part he was believed to have played in the English Navy's humiliation during the third Dutch war.

When Newton's *Principia*, substantially based on astronomical data compiled at the Royal Observatory at Greenwich, was published in 1687, Edmond Halley sent a presentation copy, printed on particularly fine paper, to James II. As Duke of York, James had (until 1673) been commander-in-chief of the English Navy, and had taken a close practical interest in the new science and astronomy as they affected navigation, weather-prediction (he owned a barometer) and the computation of tides. Halley's accompanying letter acknowledged that James had always been 'enclined to favour Mechanicall and Philosophicall discoveries'. Because he was now inevitably occupied with affairs of state, Halley excerpted for him the material of greatest interest to him: Newton's explanation of the way tides are produced by the effect of the moon's gravity. 'If by reason of the difficulty of the matter there be anything herein not sufficiently explained,' Halley concluded, he would be happy to explain it to him in person.[42] Moore's research-and-development-led Ordnance Office might well have continued to flourish had the Catholic James not been forced from the throne the following year.

Marking time

While the balance-spring watch affair rumbled on unresolved through the summer of 1675, Moore made sure the observatory project did not suffer similarly from indecision in high places. On 22 June 1675, Hooke received an order 'to direct [design and build] Observatory in Greenwich park for Sir J. More'. He viewed the site with Flamsteed at the end of the month, 'describd' the buildings on 2 July, and 'set out' the groundplan on 28 July. The foundation stone for the Greenwich Observatory was laid on the afternoon of 10 August. The building was near enough to completion to be used for viewing an eclipse at the beginning of June 1676 – it had been hoped that the King might attend, but in the event he failed to appear. Flamsteed moved in in July (but had been resident in Greenwich for some months previously, overseeing the building works, and installation of instruments).

The Ordnance Office had footed the bill for the building, but Moore personally paid to equip it. He had anticipated this from the beginning, in the invitation he extended to Flamsteed to come to London from his home in Derby: 'When you come I will consult about Instrumentes, the which I will cause made, let the charge be what it will,' he had written.[43] Moore had begun to discuss what was needed with Hooke and Flamsteed well before there was a building to put them in. The long telescopes, sextants and quadrants Moore provided were (for accounting purposes) gifts, and became Flamsteed's personal property, but the clocks, which had purpose-built wall spaces (and involved very high day-to-day maintenance costs), belonged to the observatory.

Clocks were every bit as crucial for a state-of-the-art seventeenth-century observatory's technical equipment as its telescopes and quadrants. Sir Jonas Moore lived up to his extravagant promise to Flamsteed to supply his observatory with the very best large pendulum clocks available. When Flamsteed moved into the Greenwich Observatory in the summer of 1676, two clocks by Thomas Tompion, inscribed with Moore's name as donor, were already installed in places specially designed for them in the Great Room. They had thirteen-foot (two-second) pendulums, suspended above their movements, and were often referred to as the 'spring-clock' and 'pivot-clock' to distinguish them, because of the differences in the way their pendulums were mounted. The mechanisms of these two precision clocks correspond remarkably closely to modifications to pendulums proposed by Hooke between 1669 and 1674, including the so-called dead-beat anchor escapement, to whose invention and improvement he, Tompion and Towneley all contributed.[44] In 1673

Hooke had advised Moore on how to modify one of his own pendulum clocks as Moore recorded in his published work of practical
mathematics, *Mathematical Compendium* (1674): 'And thus as the
Ingenuous Master Hooke first proposed, I have hang'd a swing by
my Clock to regulate it upon a Pin, that it may freely vibrate'; the
pin being 'placed upon the Ballance towards the back side', to take
the motion of the pendulum.

A third Great Room clock, with a six-foot pendulum, was probably supplied by Towneley and finished by Tompion. Flamsteed
personally ordered a clock with a three-foot pendulum from
Tompion for use in the sextant house. All these instruments
required frequent attention and adjustment to remain in smooth
working order, and Moore himself, having spent one hundred
guineas on the clocks (according to rumour), remained closely
involved with this throughout his life.[45] (One of the clocks started

11. Francis Place engraving of the Great Room at Greenwich
Observatory with its Tompion clocks

gaining seriously only months after it was installed, and finally stopped in November. Flamsteed blamed the lubricating oil: 'The dust mixed with the oil in the pivot holes was macerated into swarf which clogged the wheels and took off a great part of the force which the weight ought to have on the wheel which forces the pallets that move the pendulum.')[46]

Still, in spite of all Moore's commitment and effort, the observatory project also failed to produce concrete results during its chief patron's lifetime. Left to his own devices Flamsteed turned out to be a stubbornly slow worker, and a compulsive perfectionist. Moore had wanted the first tables to be issued from Greenwich by the end of 1675. Flamsteed told him that 'it might be better to forbeare till the spring that I might have more time to consider them'. By late in 1677 there was still nothing to show, and Flamsteed justified himself by protesting indignantly in a letter to Towneley that he would not 'have my immature papers and writeing forced forth of my hands and detained as if I had no interest in them'.[47] In 1679 Moore told that Royal Society that, finally, 'a quantity of [Flamsteed's astronomical measurements] were ready for the press whereupon Mr Hooke was desired to speak to Mr Martyn [the Royal Society's printer] concerning it', but in spite of a firm promise, no tables appeared. Instead, Flamsteed handed over a single copy of his measurements and calculations to date to fellow-astronomer (and Clerk at the Ordnance Office) Edward Sherburne for safe-keeping.

The matter was seemingly at an impasse. At issue was a fundamental difference of opinion between the patron and his protégé over the purposes of astronomical data-collection. Moore wanted the star-charts and tables for everyday, practical navigational purposes, and required only that they be a significant improvement on any already in existence. Flamsteed (in the words of one of his early biographers) 'resolved to reform and amend the whole system, and set a noble example for future astronomers'.[48] He was not prepared

to go to press with anything short of a complete set of data, comprehensively tabulated, with all the calculations meticulously
checked. When Moore died, later in 1679, no Greenwich tables had
yet seen the light of day.

Nor were they to appear in print for a further thirty-five years.
When Newton became President of the Royal Society in 1704 he,
too, immediately put pressure on Flamsteed to release his *History of the
Heavens* (as it was now tentatively titled) for publication. Prince
George of Denmark (Queen Anne's consort) offered to print the
work at his own expense. Newton wanted Flamsteed's data in the
public domain in order to confirm the general theory of planetary
movement proposed in his *Principia*, and above all to vindicate his
inverse square law. To publish alongside Newton's general theory
would, Newton argued, be the making of Flamsteed's own reputation:

> I am of opinion that for your observations to come abroad
> thus with a theory which you ushered into the world, and
> which by their means has been made exact, would be much
> more for their advantage and your reputation, than to keep
> them private till you die or publish them without such a theory
> to recommend them. For such theory will be a demonstration
> of their exactness, and make you readily acknowledged as the
> exactest observer that has hitherto appeared in the world. But if
> you publish them without such a theory to recommend them,
> they will only be thrown into the heap of the observations of
> former astronomers.[49]

Flamsteed was even less impressed than he had been previously by
Moore's utility arguments. He had no intention of producing his
precious observations as a mere adjunct to Newtonian grand theory.
A committed, so-called 'naïve' empirical observer of the natural
world, he believed that his data alone, properly organised and computed, would reveal its own underlying laws. Any theory of lunar
motion (for example) based on his measurements would, he argued,

be his alone: 'I call it mine, because it consists of my solar and lunar tables connected by myself, and shall own nothing of Mr. Newton's labour till he fairly owns what he has had from the Observatory.'[50]

In the end Flamsteed was duped into handing over his *magnum opus* (only two-thirds completed in his terms) by subterfuge. Edmond Halley helped put the manuscript into shape and see it through the press; the two volumes of the *History of the Heavens* came out in 1712. The single copy of the manuscript he put in Newton's hands for safe-keeping was simply passed straight to the press. In his short preface Halley offered as his assessment of Flamsteed's more than thirty years of observing at Greenwich the view that 'nothing has emerged from the Observatory to justify all the equipment and expense so that he seemed, so far, only to have worked for himself, or at any rate for a few of his friends'.[51] Flamsteed's own three-volume version of his *History* – still described by its author as incomplete – was published posthumously by his widow and two of his observatory assistants in 1725.

We should not forget, however, that Flamsteed's was not the only *magnum opus* of this period to remain uncompleted for more than thirty years. At the height of his quarrel with Newton over his refusal to 'rush out' his *History*, when challenged to justify the time it was taking to bring the work to fruition, Flamsteed replied: 'They may as well ask why St Paul's is not finished'. Begun around 1670, Sir Christopher Wren's masterpiece was finally completed in 1710–11.[52]

All at sea

Obdurate as Flamsteed was about producing his *magnum opus* before he was well and truly ready, in the 1680s he was prepared to compromise (in the spirit of his patron Sir Jonas Moore) and to produce English tables for the moons of Jupiter, relative to Greenwich as meridian, for English surveyors and navigators, matching the ones the French Astronomer Royal, Cassini, had produced for Paris. Between 1683 and 1687 he published tables for the period from the last three months of 1683 through to the end of 1688. However, Flamsteed, accustomed to making his observations under ideal conditions in his purpose-built observatory, still could not resist blaming the seamen's own fecklessness for the difficulties they had in using astronomical methods for finding longitude:

> I must confess it is some part of my design to make our more knowing Seamen ashamed of that refuge of Ignorance, their Idle and Impudent assertion *that the longitude is not to be found*, by offering them an expedient that will assuredly afford it, if their Ignorance, Sloth, Covetousness, or Ill-nature, forbid them not to make use of what is proposed.[53]

This was less than fair on the seamen. Observing the moons of Jupiter required a long telescope, held steady enough to focus on the tiny moving moon, while the observer counted the seconds before it disappeared completely behind Jupiter itself, and then kept counting till it emerged again from behind Jupiter's shadow. The likelihood of being able to do this reliably from the deck of a ship at sea was small; the likelihood of finding conditions right, and the sea consistently calm enough to do it over periods of time, remote. The French, who championed the moons of Jupiter method, did so by and large for observation work conducted on land. (In spite of the fact that Jean Richer and the two other trained observers sent by the Académie to the Cape Verde islands in 1681 were given instructions on how to make observations of the moons of Jupiter 'during Voyages', they were actually expected to make landfall before they began their observations.)

Even on land, a highly trained astronomical observer and experienced navigator using a makeshift observatory could not observe with anything like the precision of Flamsteed at Greenwich or Cassini in the Paris Observatory. During his Atlantic voyages, on the

12. Jonas Moore's *New System of Mathematics* showing navigators and astronomers with their instruments

coast of Brazil, Edmond Halley used a long telescope to observe the moons of Jupiter. In March 1699, he missed a moon's eclipse because of 'the great hight of the Planet, and want of a convenient support of my long Telescope made it impracticable'. A month later, in Barbados, he was in the middle of an observation of the ingress (disappearance) of the first moon when, 'the wind shakeing my Tube, I was willing to gett a more coverd place to observe in, that I might be more Certain, but when I again gott sight of the Planett the Satellite appeared no more'.[54]

The most onerous, physically demanding work the sheltered Flamsteed could imagine was that carried out in his own observatory — back-breaking labour, he insisted in 1710, 'labour harder than threshing'.[55] A lifelong scholar, the most taxing voyage he ever embarked upon was the one from Derby to London to take up his post as Astronomer Royal. It is not difficult to see why Flamsteed got on no better with the globe-trotting Halley than he did with the ingenious Mr Hooke.

A final word on patents

By the end of 1675 the dispute between Huygens and Hooke over priority for the balance-spring watch was petering out, although Hooke kept it going into the following year in the form of a public personal vendetta between himself and Oldenburg, conducted in the pages of their respective publications. The issue of the patent, however, dropped out of sight. By this time it was being conceded by all parties that, in spite of the shared invention of a spring-regulated balance, there was 'a vast difference' between the two watches.[56] A blanket patent covering all balance-spring watches was out of the question; and, indeed, once the concept was fully in the public domain, balance-spring watches of all kinds, ingeniously developed by skilful watch-makers, became widespread.

The next generation of watches produced further improvements to the smooth, accurate running of pocket watches. Friction at a clock's pivot points was an obvious weakness in its smooth running – lubricating oils of all kinds were tried, which tended to accumulate dust, gradually bringing the clock to a standstill (as with Tompion's clock at Greenwich).

In the early years of the eighteenth century, the Genevan mathematician Nicolas Fatio de Duillier, working in London with two French refugee watch-makers, Pierre and Jacob Debaufre, hit on the idea of piercing rubies and using them as bearings and end stones for the balance staff. Fatio had offered his jewelled watch to the French Académie, but had been rebuffed because he was a Protestant (growing Catholic intransigence had eventually driven Huygens out of France also).[57] In 1704 the three men applied for a patent for their clock-jewelling, which was initially granted but subsequently withdrawn. Their application was successfully opposed by the Clockmakers' Company on the grounds that it would be extremely damaging to the clockmaking community to patent so unspecific and generally applicable an invention. The Clockmakers expressed their support for those (like Fatio) fleeing religious persecution in France, but insisted on the damage the proposed patent would do to their own business: 'If such Watches as these Persons pretend to make should come into use, all other Watches will be undervalued, and consequently few or none of them will be made, and all the Workmen now imploy'd therein must become Servants or Tributary to these *French* Patentees or go into Foreign Parts to exercise their Trades.'

What is astonishing over these years is the sudden acceleration in clock precision, produced by the pressure to resolve longitude. The pendulum clock, which Huygens and Hooke were competing to replace, was itself a recent innovation, which had dramatically improved the precision of time-pieces less than twenty years earlier. The pendulum-driven clock had been conceived by Galileo in 1637 or earlier, and realized by Christiaan Huygens in 1656. Working with a technician clockmaker in Amsterdam, Huygens simply eliminated the balance-wheel as the regulating motion and substituted a pendulum hanging freely from a cord or wire. At a stroke, the amount the clock deviated from correct time reduced from around fifteen minutes to ten or fifteen seconds a day. The pendulum was such a simple device that not only were all upright clocks constructed with pendulums from then on but existing clocks were converted *en masse*. With that mass conversion, the conditions for precise measurement of all kinds were established, and science as we know it became a possibility.[58]

CHAPTER 5

BREAKING NEW GROUND

Making connections

THE OUTCOME OF the balance-spring watch dispute between Huygens and Hooke was a disappointment for both parties. Neither man profited (financially, or in terms of enhanced reputation), nor was the priority issue resolved to the satisfaction of either. In some ways the incident turned out to be an omen for the pursuit of the sciences in general in the period. Progress in many areas of scientific endeavour – cartography and surveying, in particular – during the last quarter of the seventeenth century and the first of the eighteenth depended, in the end, on collaboration, both between individuals and across national boundaries.

The scale of this international co-operation was remarkable, and extended well beyond the frontiers of Europe. Although this was a period of almost continuous hostilities (or at least diplomatic tension) among the three leading maritime nations (England, France and the Dutch Republic), astronomical and cartographic initiatives vital for navigation proceeded on the basis of vigorously pursued joint initiatives, and free exchanges of information, between mathematicians and astronomers from all three territories.

Within days of Charles II handing Sir Jonas Moore the royal war-rant for building the Greenwich Observatory on 22 June 1675 the precocious Edmond Halley joined Moore, Flamsteed and Hooke in Moore's rooms at the Tower of London, to observe a total eclipse of the moon. According to the report of the observations which Flamsteed made to the Royal Society, Halley 'assisted carefully with many of them'.[1] A few days later Halley accompanied Hooke and Flamsteed to Greenwich to inspect the site of the future observatory.

The English astronomers did not undertake their observation of the eclipse alone. In France, simultaneous observation of predictable celestial events (total lunar eclipses, eclipses of the moons of Jupiter, transits of Mercury or Venus across the sun) was well established as the way to determine longitude accurately on land (total lunar eclipses, unlike partial ones, are visible wherever it is night when the eclipse takes place). Cassini had already collaborated with James Gregory, Regius Professor of Mathematics at St Andrew's University in Scotland, to observe an eclipse of the moon in 1674. Gregory was the first incumbent in a Regius Chair established in 1668 by Charles II at the personal request of Sir Robert Moray. In 1674 his observatory had been in use for less than a year.

Halley and Flamsteed celebrated London's emerging importance as a centre for astronomical observation exactly as Gregory had. At precisely the moment they were observing the lunar eclipse in London, the French Astronomer Royal, Cassini, and his assistants, were doing the same thing, by prior arrangement, at the Paris Observatory. Immediately afterwards they exchanged the data col-lected. By comparing the times recorded, the two collaborating teams were able to obtain precise measurements for the relative lon-gitudes of their two locations.[2] Halley, Hooke and Flamsteed's subsequent trip to the Greenwich site may well have involved dis-tance calculations (of the kind Hooke had practised with Wilkins, using his reflecting quadrant, ten years earlier)[3] so that they could correct the observations of the lunar eclipse they had made at the

Tower, to correspond to the Greenwich location. This would ensure that, for future reference, the two sets of observations would accurately relate the longitude of one national observatory to the other.

Although it took considerably longer for his data to reach home, the 26 June 1675 lunar eclipse was also being observed by Halley's close friend Charles Boucher, who had gone to Jamaica on an astronomical fact-finding journey, which probably inspired Halley to make his own to St Helena the following year. Boucher's letter, containing his data for the June 1675 eclipse, was sent to Halley, and passed on to Flamsteed in May 1676. The letter informed Halley (and via Halley, Flamsteed) that conditions in Jamaica were excellent: 'Tho the heavens be cloudy and it raine all day, yet the clouds set immediately after the sun, and the nights are allwayes serene.' Boucher also informed his fellow-astronomers that he had been shipwrecked on his outward journey, and lost almost all the tables and instruments he had taken. Flamsteed sent him replacements early the following year.[4] From the observations Boucher had made, Flamsteed calculated that the difference in meridians between Greenwich and Jamaica shown on 'the usuall Mapps' was out by as much as five degrees.

When Halley decided to go south of the equator to St Helena, the most southerly of England's possessions, in October 1676, Cassini immediately proposed an Anglo-French joint programme of astronomical research. The Paris Observatory had already organised simultaneous observations of lunar eclipses and eclipses of the moons of Jupiter with observers in French Canada and at Cayenne in French Guiana. Cayenne, however, although on the other side of the Atlantic Ocean, is still north of the equator. Halley would observe designated eclipses from his makeshift observatory on St Helena; Flamsteed and Cassini would do the same at Greenwich and the Paris Observatory respectively.

As the basis for these prearranged simultaneous observations, Cassini sent Halley a copy of his most up-to-date data on lunar

ASTRONOMIA CAROLINA,

With Exact and most easy

TABLES and RULES

FOR THE

Calculation of ECLIPSES.

By THO. STREET.

The Third Edition, Corrected.

TO WHICH

Is Added a Series of Observations

ON THE

PLANETS,

Chiefly of the

MOON,

MADE NEAR

LONDON:

With a Sextant of near Six Foot Radius; in Order to find out the LUNAR THEORY a Posteriori. Being a Proposal how to find the LONGITUDE, &c.

By Dr. EDMUND HALEY.

LONDON:

Printed for S. BRISCOE, and R. SMITH, at the Royal-Exchange Cornhil, MDCCXVI.

1. Title page of Halley's revised edition of Streete's *Astronomia Carolina*

eclipses and on the moons of Jupiter to take with him. Halley also took the equivalent English compilation, Thomas Streete's *Astronomia Carolina* (1664) (less accurate, but more conveniently calibrated taking London as its zero meridian). Armed with his tables, Halley could co-ordinate his astronomical observations with his fellow observers back in Europe. Simultaneous measurements could then be used to help solve long-standing theoretical problems in astronomy (Cassini suggested that Halley's lunar eclipse observations might help fix a value for the distance of the moon from the earth). They could also improve the accuracy of the tables of prediction themselves. (In 1710 Halley published a revised edition of Streete's *Astronomia Carolina,* based on the measurements he had collected following the St Helena voyage.)

So although the official purpose of Halley's trip – the one used to get the East India Company's sponsorship – was to map the stars in the southern hemisphere, he had also promised Cassini in Paris, and Flamsteed and Streete in London, that he would observe a total eclipse of the moon on 6 May 1677, and a transit of Mercury on 28 October 1677 (old style). Indeed, it was in order to be sure of reaching St Helena in time for these that he left Oxford in haste, without collecting his BA degree, in late October 1676.[5] Streete and Hooke duly attempted to view the transit of Mercury, in London simultaneously with Halley, and Towneley in the north of England tried too. However, as Hooke noted that day in his diary, 'The eclipse of [the sun] by [Mercury] appeared not' – apparently because bad weather obscured the sky right across Britain.[6]

In spite of the distances involved, the travellers got their observational data home by the first reliable carrier they could find. By early the following year Moore and Flamsteed had sent Halley's data for the two astronomical events to Cassini in Paris, Jean Charles Gallet in Avignon, Hevelius in Gdansk, and Charles Boucher in Jamaica. They all sent their own data to London in return. Sadly, of all the co-ordinated centres of observation involved, the new

Greenwich Observatory turned out to yield the most disappointing results. In a letter to Boucher, Flamsteed admitted that he had been unable to contribute much in the way of observations, because 'our weather has been so Cloudy this last 12 Months'.[7]

From the French point of view, however, the St Helena expedition was an unqualified success. Thus it was that at the very beginning of 1681 Halley (not yet twenty-five) arrived in Paris on a semi-official visit, to a hero's welcome. Cassini took him to a meeting of the Académie Royale des Sciences (which did not normally admit non-members), where Halley reported on 'the observations he made of Mercury crossing the face of the Sun'.[8] *En route* for Paris Halley got his first sight of the spectacular 1680–81 comet shortly before dawn on 18 December (new style), at Calais, after a stormy forty-hour crossing. Once he reached Paris, he joined Cassini at the Royal Observatory to observe the comet whenever it was visible. Cassini reported that 'that excellent astronomer from the Royal Society in England was present at most of my observations' and that Halley had computed many of the comet's positions.[9] Meanwhile, Halley wrote to Hooke in London promising to send the Paris comet data to London, and requesting that Hooke 'do me the favour to let me know what has been observed in England'. When Cassini's pamphlet on the comet appeared later that year it included Flamsteed's Greenwich observations of the comets, supplied by Flamsteed via Halley. In late January Halley sent Flamsteed more data on the comet compiled in Rome and Avignon.

On 29 August 1681 (new style), Halley was in Avignon, on his way to Italy. There he observed a total eclipse of the moon with Jean Charles Gallet; they sent their observation to Cassini and Flamsteed. By comparing the values obtained, Flamsteed derived the difference in longitude between Greenwich and Avignon. Passing through Paris again on his way home, Halley met Cassini again in January 1682, and gave him the values for the latitudes he had measured,

2. Cassini's notes on Halley's observations of a lunar eclipse

using a small quadrant, at more than a dozen places he had visited during his tour.[10]

Sharing experience and astronomical information in this way could occasionally lead to a fundamental theoretical breakthrough. During his Paris visit, Halley was also given access to the data Jean

Richer had just brought back from an astronomical data-collecting voyage to Cayenne in French Guiana and Gorée in the Cape Verde islands. At Cayenne Richer had found that the pendulum clock he had set exactly in Paris 'went too slow as every day to lose two minutes and a half, and after his clock had stood [stopped] and went again it lost two and a half minutes as before' (as Picard had already noted at the same location ten years earlier). Halley had encountered the same phenomenon on St Helena. He advised Richer and Cassini that to ensure that their clocks would run accurately at Cayenne they needed to shorten their pendulums by 0.2 centimetres.

Halley later passed all this information to Newton. St Helena is somewhat farther south of the equator than Cayenne is north. At both places it had been found necessary to shorten the pendulum of a clock to make it run true. Six years later, Newton, in his *Principia*, concluded that near the equator gravity is less because the earth is not perfectly spherical, but rather an oblate spheroid (compressed at the poles; bulging at the equator).[11]

Richer's observations of pendulum lengths on Cayenne give every indication that he was as skilful with clocks as with his astronomical instruments. Huygens, however, held a different view. On an earlier scientific voyage to Cayenne, in 1670, Richer was put in charge of the sea-trials for two of Huygens's experimental pendulum clocks. At first the clocks seemed to have performed well, even during a storm. Later on, however, they proved less reliable. Huygens claimed, indignantly, that Richer had become seasick and neglected the clocks entirely, but we should probably take this with a pinch of salt. Huygens failed to produce his clock in time to be tested on Richer's next voyage to Cayenne in 1672 (the first voyage on which it was noted that a pendulum clock ran consistently slow close to the equator).[12]

During Halley's stay in Paris, Cassini in his turn shared material and ideas with him, which would influence Halley's own later ideas on the periodic return of orbiting comets. On 29 May 1681 (new style) Halley wrote to Hooke from Saumur:

Monsieur Cassini did me the favour to give me his booke on
the Comett just as I was goeing out of towne; he besides the
Observations thereof, which he made till the 18 of March new
stile, has given a theory of its Motion which is, that this
Comet was the same with that that appeared to Tycho Anno
1577, that it performes its revolution in a great Circle
including the earth, as likewise to that of April 1665. I know
you will with difficulty Embrace this Notion of his, but at the
same Tyme tis very remarkable that 3 Cometts should so
exactly trace the same path in the heavens with the same
degrees of velocity.[13]

Cassini was wrong about the equivalences between comets, but
right about their returning orbits. The months spent with Cassini
introduced Halley to ideas well in advance of those in circulation at
home during the same period. At almost the same moment in
England, Newton was still insisting to Flamsteed that comets traced
linear paths, and that the suggestion that the comet observed at the
end of 1680 and the one observed at the beginning of 1681 were
one and the same was 'to make that one paradoxical'.[14]

In spite of Sir Jonas Moore's anxiety that the Greenwich
Observatory might be perceived as generating little astronomical
material of practical consequence, Flamsteed and Halley did in fact
produce and publish some short-term tangible results before 1700.
During the forty years while Flamsteed was diligently pursuing his
long-term star-mapping project, however, these tended not to be
specifically English initiatives, but rather to support those from
Cassini's French Observatory. Each year from 1683 to 1688
Flamsteed published tables for eclipses of the moons of Jupiter for
English use, calculated relative to the Greenwich meridian and Julian
calendar. In 1693, when Cassini reissued his much improved tables
(now for the first moon only), Halley computed the corrections
needed to bring these in line with the Greenwich meridian and
Julian calendar, and published an English edition.

[413]

A Catalogue *of the* Visible Eclipses *of* ♄ Satellits, *shewing the apparent times of their* Ingresses *into* ♃ *shadow and* Emersions, *from it under the* Meridian *of the* Observatory *in the year* 1684. *Calculated from new* Tables *of their* Motions. *by* John Flamsteed *M. R & R. S. S.*

1684

Jan.	h	'			Feb.	h	'			Mar.	h	'			Apr.	h	'		
♂ 1	13-03	I	*	i	♃ 1	5 22	2		i	♄ 1	19-24	I		e	♂ 1	18-23	2		e
☿ 2	23-30	3		i	♄ 2	9-47	I	*	i	☽ 3	13-53	I	*	e	☿ 2	16-09	I		e
♃ 3	7-31	I		i	☽ 4	3-,6	I		i	♂ 4	7-54	2	*	e	♀ 4	10-38	I	*	e
	19-09	2		i		19-39	2		i	☿ 5	8-22	I	*	e	♄ 5	6-26	3		e
♄ 5	1-58	I		i	♂ 5	22-22	I		i	♃ 7	2-51	I		e		7-42	2	*	e
☉ 6	12-29	4	*	i	♃ 7	6-53	I	*	i		14-24	3	*	e	☉ 6	5-07	I		e
	15-39	4	*	e		19-13	3		i		21-1	2		e	☽ 7	23-36	I		e
	20-26	I		i	♀ 8	7-57	2	*	i	♄ 8	21-20	I		e	♂ 8	21-01	2		e
☽ 7	8 25	2		i	♄ 9	0-26	4		i	☽ 10	15-5c	I	*	e	☿ 9	18-05	I		e
♂ 8	14-54	I	*	i		3-19	4		e	♂ 11	10-30	2	*	e	♀ 11	12-34	I	*	e
♃ 10	3-36	3		i		11-21	I	*	i	☿ 12	11-19	I	*	e	♄ 12	10-19	2	*	e
	10-22	I	*	i	☽ 11	5-50	I		i	♃ 13	15-13	4	*	e		10-26	3	*	e
	21-42	2		i		21-15	2		i	♀ 14	4-48	I		e	☉ 13	7-03	I		e
♄ 12	4-49	I		i	☿ 13	0-19	I		i		18-25	3		e	♂ 15	1-32	I		e
☉ 13	22-17	I		i	♃ 14	18-47	I		i		23-49	2		e		23-38	2		e
☽ 14	10-58	2	*	i		23-13	3		i	♄ 15	23-17	I		e	☿ 16	0-54	4		i
♂ 15	16-46	I	*	i	♀ 15	10-34	2	*	i	☽ 17	17-4c	I		e		3-06	4		e
♃ 17	7-22	3		i	♄ 16	13-16	I	*	i	♂ 18	13-09	2	*	e		20-01	I		e
	11-13	I	*	i	☽ 18	7-45	I	*	i	☿ 19	12-15	I	*	e	♀ 18	14-30	I	*	e
♀ 18	0-16	2		i		23-52	2		i	♀ 21	6-44	I	*	e	♄ 19	12-46	2		e
♄ 19	5-41	I		i	☿ 20	2-14	I		i		22-26	3		e		14-26	3	*	e
☽ 21	0-10	I		i	♃ 21	20-42	I		i	♄ 22	2-28	2		e	☉ 20	18-59	I		e
	13-32	2	*	i	♀ 22	3-1	3		i	☉ 23	1-13	I		e	♂ 22	3-28	I		e
♂ 22	18-38	I	*	i		13-11	2	*	i	☽ 24	19-43	I		e	☿ 23	2-15	2		e
☿ 23	6-27	4		i	♄ 23	15-12	I	*	i	♂ 25	15-46	2	*	e		21-57	I		e
	9 26	4	*	e	☽ 25	9-41	I	*	i	☿ 26	14-12	I	*	e	♀ 25	16-26	I		e
♃ 24	11-19	3	*	i	♂ 26	☉♃				♀ 28	8-41	I	*	e	♄ 26	15-33	2		e
	13-06	I	*	i	☿ 27	6-26	I		e	♄ 29	2-26	3		e		18-26	3		e
♀ 25	2-47	2		i	♃ 29	0-55	I		e		5-05	2		e	☉ 27	10-55	I	*	e
♄ 26	7-34	I		i		10-24	3	*	e	☉ 30	3-10	I		e	♂ 29	5-24	I		e
☽ 28	2-0	I		i		18-35	2		e		9-09	4	*	e	30	4-51	2		e
	16-05	2	*	i						☽ 31	21-39	I		e		23-53	I		e
♂ 29	21-30	I		i															
♃ 31	14-5	I	*	i															
	15-16	3	*	i															

May

3. Flamsteed's predictions of eclipses of the moons of Jupiter for 1684

Giving shape to the globe

Cassini was confident that his increasingly precise tables for the movements of the moons of Jupiter were the right method to produce exact global positioning because their worth had been well proven on land. French astronomers' efforts during this period were squarely focused on a scientific goal consistent with Louis XIV's expansionist imperial dreams. The top priority for the Paris Observatory was to use astronomical methods to produce a definitive world map. When the King tempted Cassini to Paris in 1669 (on what both Cassini and his University of Bologna employers believed was to be only a brief visit) it was in order to acquire his recently published tables for the eclipses of the moons of Jupiter to apply to the large-scale enterprise of accurately mapping first France and then the whole of the known globe.

Good maps were a strategic necessity for Louis XIV's expansionist military operations, and the prime reason behind his agreeing to generous funding for the Académie Royale des Sciences. The problem of mapping France was discussed at the first meeting of the Académie. In May 1668, Jean Baptiste Colbert, Minister of Finance and of the Navy, sent a message to the Académie's weekly meeting that they should direct their efforts without delay towards making maps of France more accurate than those currently available.[15] Once it was clear that Cassini's method yielded remarkably good measurements for mapping purposes, Colbert was authorised to offer him a handsome pension, and he was set the project of drawing up a detailed map of France in multiple sheets, the entire exercise to be done using consistent methods, to uniform standards of accuracy, and employing standard symbols.[16]

The first step in such an undertaking was to establish a basic scale. In 1669 Jean Picard set out to measure the arc of the meridian at Paris (the north–south distance from one degree of latitude

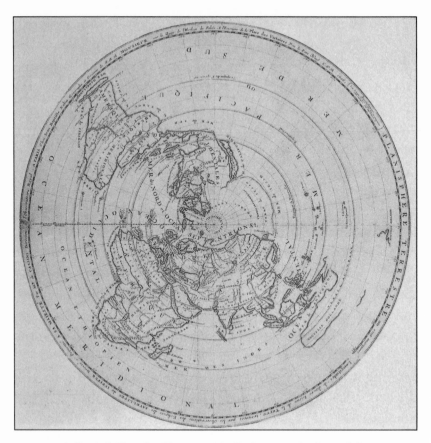

4. Cassini's *Planisphere Terrestre*, originally drawn on the floor of the
Paris Observatory

to the next) in order to determine the size of the earth. He used a
triangulation method for establishing distance (proposed over a
century earlier by Gemma Frisius). Taking as base points the
Pavilion at Malvoisine near Paris and the clock tower in Sourdon
near Amiens, thirteen large triangles were surveyed to give the precise distance between these reference points. Observations of
Jupiter's moons were made using a quadrant of his own design
(with telescopic sight and filar micrometer), three long telescopes
and two pendulum clocks, one with a pendulum beating once per

second, the other clock beating every half second. The results of the survey gave the diameter of the earth as 57,060 toises (about 12,554 kilometres, a good result compared with the equatorial diameter of 12,756 kilometres used today). Ironically, the exactness with which Picard could propose his calculations (made using spherical trigonometry, then projected on to the plane for mapping purposes) depended on his unshakeable conviction that the earth was a perfect sphere – a working assumption that was experimentally disproved by measurements taken in the course of the French mapping project itself.

Cassini proceeded to collate celestial observations provided by informants (including Halley) from around the world to create an accurate map of the known world. He laid out a planisphere (world map), using an azimuthal projection with the North Pole at the centre, on the floor of the Paris Observatory. Although the projection chosen greatly distorted land shapes it produced latitudes as concentric circles centred on the North Pole, and longitude as equally spaced radii. A cord was attached to the polar centre with a movable pointer on it. If the pointer was set to the correct latitude and the cord rotated to the correct longitude, the position of any desired place on the globe could be given precisely – an exercise performed regularly for the entertainment of members of Louis XIV's court.[17]

After the death of Cassini senior, Jacques Cassini de Thury succeeded his father as the head of the Paris Observatory and continued the survey of France. The topographical map of France – the most ambitious mapping project ever undertaken – was finally completed by de Thury's son in 1789, having cost the nation 700,000 livres. The sheets, when assembled, formed a map 33 feet by 34 feet in size.[18] There is some irony in the fact that the project, carried out by three generations of a single Italian immigrant family, which rose to the rank of minor nobility in the process, was presented to the National Assembly of the French Revolutionary

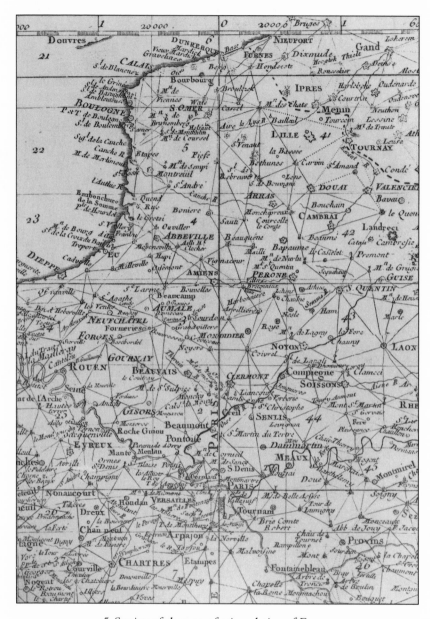

5. Section of the map of triangulation of France

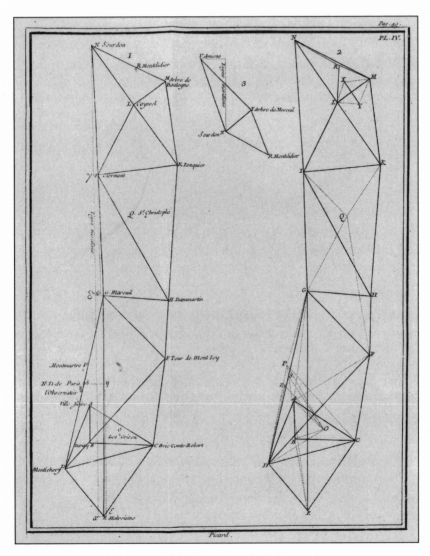

6. Diagram of Cassini's method of triangulation

Government by Jacques-Domenique Cassini. The National Assembly immediately recognised the map as a military and political asset of the utmost importance, and confiscated it.[19]

One of Cassini's early initiatives in support of the mapping project was to send Picard to Uraniborg to recalculate the longitude at the observatory of the great Danish astronomer Tycho Brahe. Picard reported a fifteen-minute discrepancy between Cassini's value for the predicted disappearance of the first satellite of Jupiter and his observed time of actual ingress. Over the next five years, teams of observers reported consistent errors of the same order. On the basis of this accumulated evidence, one of the Paris Observatory team, Ole Rømer, proposed in 1676 that the problem had to do with the finite speed at which light itself travelled: light from a celestial event took appreciably longer to reach an observer when the planet (say) under observation was at its greatest orbital distance from the earth. (Cassini himself never accepted Rømer's explanation; Newton and Halley, on the other hand, did.)[20]

In 1683, Colbert died, and supervision of the Paris Observatory's expenditure was taken over by the head of the French military, Michel François Le Tellier, Marquis de Louvois. Louvois considerably curtailed the Académie's funding, but remained committed to the mapping project. Instead of sending out special scientific expeditions to the Far East, Louvois had the Académie train Jesuit missionaries in mathematics and astronomy. The Jesuits were then sent off on subsidised missions, lavishly equipped with books, charts and scientific instruments, and with instructions to send back all kinds of technical data: measurements of latitude and longitude, astronomical observations, reports about flora and fauna, calendars, alphabets and numerical systems.[21]

In 1685, for example, the Académie heavily subsidised Father Guy de Tachard and six Jesuit colleagues on a missionary voyage to China. Rather like the additional 'pay-loads' space-shuttle astronauts carry on missions today, the members of Tachard's expedition to convert the heathen were expected to repay the investment by carrying out experiments for the Paris Académie along the way. They took with them thermometers, barometers and air-pumps for the purpose, together with careful instructions on how to set up their experiments. As Tachard explained in his published account of the

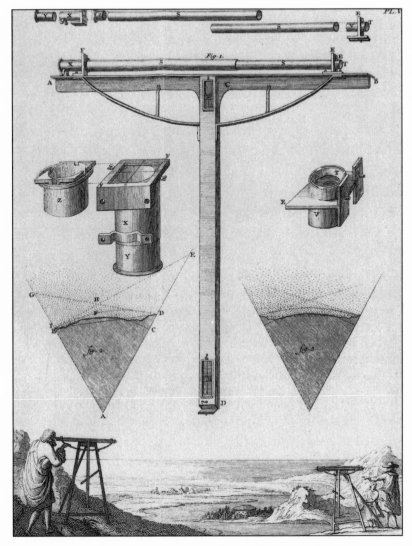

7. Surveying instruments and triangulation in action

voyage, the prime scientific object of this particular expedition was
astronomical and cartographic:

> Since the time the King settled a Royal Academie at *Paris*, for
> improving Arts and Sciences within his Kingdom, the Members

that compose it, having not hit upon any means more proper
for accomplishing that Design, than the sending out of
Learned Men to make Observations in foreign Countries,
whereby they might correct the Geographical Maps, facilitate
Navigation, and raise Astronomy to its Perfection.[22]

Cassini's observatory provided Tachard with a considerable quantity of high-grade scientific instrumentation and equipment,
including long, high-resolution telescopes (among them 'an
Excellent Telescope twelve foot long, of the late Monsieur le Bas')
and two pendulum clocks made by Isaac Thuret (formerly
Huygens's instrument-maker). Naturally they also carried Cassini's
tables for the eclipses of the moons of Jupiter, and full instructions
on how to use them.

They began their astronomical observations well before they got
to China. Having reached the Cape of Good Hope, Tachard and his
colleagues were pleasantly surprised to be well received by the Dutch
Commissioner at the Cape, Simon van der Stel (in 1685 Louis
XIV rescinded the Edict of Nantes and expelled all Huguenots; the
Catholic fathers hardly expected a welcome from the representative
of a Protestant power). Tachard told van der Stel that they were
equipped to recalculate the longitude of the Cape itself:

> We explained to him the new way of observing by the *Satellites*
> of *Jupiter*, of which the Learned *Monsieur Cassini* hath made so
> good Tables. I added that thereby we would render a very
> considerable Service to their Pilots, by giving them the certain
> Longitude of the Cape of *Good-hope*, which they only guessed at
> by their reckoning, a very doubtful way that many times
> deceived them, and that very considerably too.[23]

Van der Stel was enthusiastic, and set Tachard and his colleagues
up in a 'pavillion' within the grounds of the Dutch East India
Company's botanical garden (established at the Cape by Jan van

Riebeeck in 1652) 'which with the Dutch I shall call our Observatory':

> There is a great Pile of Building built at the Entry into the Garden, where the Companies Slaves live. It contains a Pavilion open every way, betwixt two Tarasses paved with Brick, and railed about; the one looking towards the North, and the other to the South. This Pavilion seemed to be purposely made for our Design: For on the one side we discovered the North, the View whereof was absolutely necessary to us, because it is the South in relation to that Country.[24]

On the basis of the observations of the moons of Jupiter they made from their makeshift observatory, Tachard proposed relocating the Cape on navigational charts, since 'the Charts are defective, and make the Cape more Easterly by near three degrees than in reality it is'.[25] The longitude of the Cape was a vexed issue, the more so because precision in locating its southernmost tip was a matter of such importance to sea-captains and their navigators. Halley, reviewing Tachard's book for the Royal Society's *Philosophical Transactions*, took issue with the Jesuits, casting doubt on the accuracy of their measurements and the reliability of their clocks.[26]

The clocks Tachard was travelling with were precision pendulum clocks built by Isaac Thuret, the clockmaker who had worked with Christiaan Huygens on his prototype balance-spring watch in 1675, and had himself claimed the French patent for it. By one of those small twists of fate characteristic of the pursuit of scientific excellence in the period, Tachard's act of astronomical generosity towards the Dutch East India Company at the Cape of Good Hope brought Thuret and Huygens once more into competition.

In October 1682 Huygens, back in The Hague, had obtained funding from the Dutch East India Company to work on a pendulum-based marine timekeeper with the Dutch clockmaker Johannes van Ceulen. The clock was not ready for sea trials until August

8. Cape Observatory from Tachard's *Voyage to Siam*. Note the clock and
the telescope

1675, when Huygens and Johannes de Graaf took it out on the
Zuiderzee (it performed tolerably well). Huygens also trained a
seaman, Thomas Helder, to look after the clock on a longer test-
voyage. In June 1686, less than a year after Tachard, de Graaf and
Helder set out for the Cape on the Dutch East India Company's
ship *Alcmaer*, with Huygens's clock on board. The Company was not
impressed, however, with the measurements achieved with Huygens's
clock, on this or a second trial voyage to the Cape of 1690. It
withdrew its funding, and the project ceased. Before 1700, the
moons of Jupiter method for establishing longitude that Tachard
was using looked a far better commercial bet than Huygens's over-
sensitive chronometer.[27] So on this occasion, at least, a Thuret clock
had outperformed a Huygens one.

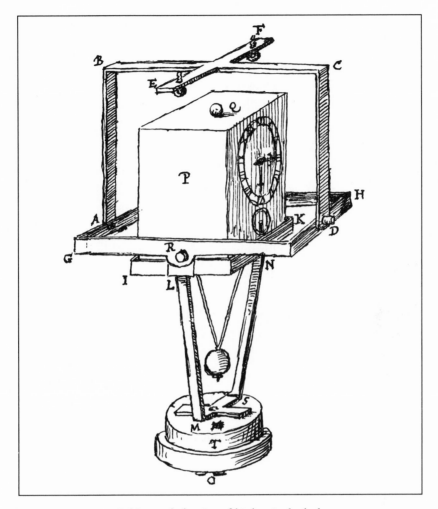

9. Huygens's drawing of his longitude clock

Ironically, in spite of the interdependency of their two fields, cartographers and theoretical astronomers ultimately came to blows over the interpretation of errors in the observational data collected. For the theoreticians, consistent errors confirmed grand theories: time errors shown by the first of Jupiter's moons encouraged Rømer to propose a finite speed for light; Picard and Richer's pendulum observations confirmed the oblate spheroid shape of the earth proposed by Newton. For the cartographers, however, it was of fundamental importance that they be able to adhere to the geometrical assumptions made when they began their ambitious mapping project. Thus Picard and Cassini resolutely stuck with the perfectly spherical earth for cartographic purposes, and Cassini preferred an *ad hoc* angular correction to rescue the data in his tables for the moons of Jupiter. It was not until 1740, following his expedition to measure the length of an arc of the meridian at Tornio in Lapland, that Louis Moreau de Maupertuis finally persuaded Cassini's son, Cassini de Thury, to admit that the earth was almost certainly an oblate spheroid, compressed at the poles. Voltaire remarked sarcastically that Cassini de Thury had finally confirmed by observation 'in some of the world's most boring places', 'what Newton knew without ever leaving home'.[28]

When Picard calculated his value for the length of an arc of the meridian at Paris in 1669, Charles II sent a message to the Royal Society, via John Wilkins, that he wished the same to be done in London. Neither the King nor the Society, however, was prepared to put up any money for the project. Seventeen years later, in July 1686, Edmond Halley (as one of his first projects as salaried clerk to the Royal Society) proposed replicating Picard's method of finding the arc of the meridian by triangulation. On the very day that the Society agreed to license Newton's *Principia* for publication they also agreed

> That the treasurer, to encourage the measuring of a degree of the earth, do give to Mr Halley fifty pounds or fifty copies of Willoughby's *History of Fishes* when he shall have measured a degree to the satisfaction of Sir Christopher Wren, the president, and Sir John Hoskyns.[29]

Fifty pounds, Halley explained, would not even cover his costs. Nothing came of Halley's project, and the triangulation of England had to wait until the eighteenth century, when General William

Roy set up the Ordnance Survey. Roy, an accomplished surveyor, engineer and archaeologist, had produced a military map of Scotland during the 1745 Scottish rebellion. Recognising the advantage good maps gave an army, he proposed a national survey of Britain along the same lines. (A national organisation responsible for surveying and mapping the country was not in fact established until after Roy's death in 1790.)

In striking contrast to the French, cartography in England continued to be privately funded, either by individuals, or by organisations like the East India Company whose business depended on accurate maps, until the end of the seventeenth century. Even military maps were privately sponsored – when Pepys eventually lost his job at the Navy Office he took his considerable collection of maps with him.

Beyond the help enlisted by the Cassini family to assist their French mapping project, there were moments when the observatories in London and Paris were drawn into collaboration for straightforwardly pragmatic reasons. In spite of relatively cordial relations, extending to frequent sharing of data, the two observatories continued to use separate zero meridians, one at Greenwich and one at Marly. Worse still, the English value for the difference in latitude and longitude between Greenwich and Marly did not agree with the French one. The latitude was out by fifteen seconds, the longitude by nearly eleven seconds. In 1783, Cassini de Thury suggested that the English astronomers should make a trigonometric survey from Greenwich to Dover, while the French did the same from Marly to Calais. The two observatories would then collaborate on a triangulated connection, across the Channel, between them. This part of the project was carried out at night, using cliff-top lights and theodolites, in 1787. William Roy oversaw the English end of operations.[30]

Investing in 'terra incognita'

As the involvement of the English and Dutch East India Companies makes clear, map-making and commercial interests inevitably went strenuously hand in hand. The person with the best maps was understood to have the most reliable access to sought-after overseas commodities and distant markets. Beyond Europe, cartographic expeditions (which, as we have seen, doubled up with astronomical expeditions, since the two data-gathering activities use the same technical instruments and are closely related) followed the established trade routes of the major trading Companies to their farthest locations.

Gorée and Cayenne, the two places where Richer and his team of observers chose to take their astronomical observations in order to fix longitude, were the two end-points on the thriving French commercial slave route. Slaves captured on the African mainland were taken to a holding centre on Gorée, off the coast of Senegal. From there they were shipped to Cayenne in French Guyana. Halley charted the southern stars from St Helena, south of the equator in the Atlantic, a stopping place for ships of the East India Company, seized from the Dutch because it was critical as a location for taking provisions and water on board *en route* for the Cape of Good Hope (where the Dutch East India Company had established a sizeable scientific research presence), and thence to the lucrative markets in drugs and spices to the East.

Joan Blaeu, publisher of the greatest of the seventeenth-century printed compilation of detailed maps, the *Grand Atlas* or *Atlas Maior*, was also employed as a chartmaker for the Dutch East India Company. In spite of his commitment to producing the best possible maps, wall-charts and atlases in mass-produced volumes, Blaeu nevertheless carefully kept 'classified' a good deal of material concerning coastlines and harbours in the Far East. In the 1660s, only his manuscript maps designed for the Company's navigators incorporated all the most up-to-date material he had in his possession about the detail of the Asian coastline. The maps of Asia in his printed atlases were mainly reprinted from earlier plates, withholding the material Blaeu had compiled for his trade employers.[31]

Nor was Blaeu's an 'armchair' involvement with the speculative venture side of cartography and sea-travel. He also speculated financially in the new North American markets himself, taking investment advantage of the privileged information on routes he had been involved in charting. A 1663 contract survives in which Blaeu and four colleagues undertook to engage in 'trade and cultivation in the islands of Virginia', and to supply the necessary slaves to work their plantations.[32]

At the end-points of these exotic trade routes, the most scientific and precise maps were in great demand as gifts for the local dignitaries whose goodwill was essential to the Europeans with territorial

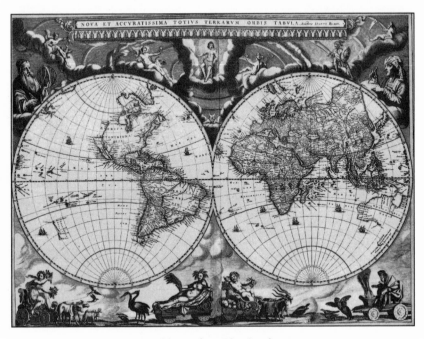

10. World map from Blaeu's *Atlas Maior*

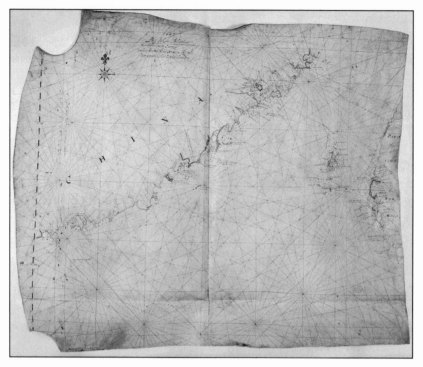

11. Manuscript map of the China Sea, by Blaeu

or commercial interests in the region. Top quality scientific instruments were also increasingly sought after, as the technology essential for local developments in astronomy and cartography to match those of Western competitors. When Tachard set out for China his considerable collection of scientific instruments included some which he was to leave 'in the Observatory of *Peguin*'.[33] Copies of Blaeu's *Atlas Maior* were produced with Turkish text, for presentation to Ottoman officials.[34]

In London, the buoyant market in atlases drew the indefatigable Robert Hooke into cartography, on top of his many other scientific pursuits. When Hooke became part of the surveying team for the rebuilding of London, following the Great Fire of 1666, he came into contact with John Ogilby and his young assistant

William Morgan, who had been appointed 'sworn viewers' by the Corporation of London, to re-establish property lines obliterated by the fire. The elderly Ogilby was a printer and bookseller, but the fire had destroyed his entire establishment and stock in Fleet Street. To relaunch his business, Ogilby undertook 'a new model of the universe, an English Atlas', for which ambitious work he obtained a licence in November 1669, with copyright for fifteen years. Both Hooke and John Aubrey became involved in the production of the 'Britannia' volume of this five-volume undertaking, providing advice and assistance with the actual mapping, and (in Hooke's case) designing technical instruments to facilitate the surveying.[35]

The project was a financial disaster, whose expenses rapidly outstripped the funding available, in spite of valiant efforts on Hooke's own part to drum up influential support among his coffee-house and Royal Society acquaintances. It eventually collapsed altogether, leaving large amounts of partially completed work, and equally large numbers of unpaid creditors, including the cartographers themselves. All that was actually published was a 'Map of London', which appeared in the New Year of 1677, signed by Morgan alone (Ogilby had died the previous year).

On 28 March 1678, another London bookseller, Moses Pitt, presented his proposals for an 'English Atlas' at a meeting of the Royal Society. Hooke and Wren had together engineered Pitt's Royal Society appearance. A committee that included Hooke was appointed to look into the feasibility of this new atlas proposal. Their decision was to reuse the plates from Blaeu's *Atlas Maior*, as already updated by the Dutch cartographic printer Jansson. (Publication of the hugely popular and financially successful Blaeu atlas itself had ceased in 1672 when the plates were destroyed in a fire at Blaeu's printing-house; Blaeu himself died shortly afterwards.) Hooke reported in his diary that it had been 'resolved about using the Dutch maps but reducing them'. The committee further advised that the atlas should be published in eight volumes, containing 202 maps with appropriate tables of names and descriptions.[36]

Pitt's atlas fared no better than Ogilby's: it was undersubscribed, and never completed. The decision to reuse old maps (however exquisitely engraved and presented) was bound to result in a botch, as Hooke pointed out on several occasions. Besides, this basically commercial enterprise lacked the kind of support in terms of money and manpower that had produced the great Dutch maps and was now producing quality French ones. In 1685 Moses Pitt was arrested for debts arising from another failed business venture, and confined to the Fleet prison. Hooke had washed his hands of the whole project long before; his last diary entries concerning Pitt refer to him as 'that rascall'.[37] (After Hooke's death, Sloane acquired some Blaeu charts, and three volumes of maps from the printed atlas of Johann Jansson which were 'intended to be the foundation of Pitt's English Atlas'.)[38]

Claiming, reclaiming and salvage

While around the globe astronomical observers and cartographers joined forces with the joint stock companies to their mutual advantage, closer to home the productive partnership was one between various types of group interests and the new technological sciences, in areas connected with drainage, land reclamation and fortification. Once again these involved international collaborations, but whereas France was the leading power in cartography, the Dutch Republic led the field in land reclamation.

On 20 May 1662, Charles II of England married the Portuguese princess, Catherine of Braganza, and received the port of Tangier (across the water from Cadiz on the southern side of the Straits of Gibraltar) as part of her dowry. In the same year, Edward Montague, first Earl of Sandwich, led an expedition to take control of England's newest territorial possession, installing a garrison and the first English governor. Sandwich personally invested in property there and over the next decade remained one of the staunchest supporters of Tangier's strategic importance, and the desirability of England's sustaining a presence there.[39]

Actually, the strategic importance of Tangier was never in any doubt; the problem was the port's vulnerability. Sir John Lawson told the King that 'if it were in the hands of the Hollanders, they would quickly make a Mole, which they might easily do; and they would keep the place against all the World, and give the law to all the trade of the Mediterranean'. It was quickly agreed that what the port needed was indeed a Mole – an enclosed, fortified harbour on the Dutch model.

The team sent out to Tangier early in 1663 to design and draw up plans for the Mole included Sir Jonas Moore. In order to qualify for a major role in the project, nominees had to have had extensive previous surveying experience. Wren, already judged a surveyor of stature for the structural innovativeness of his Sheldonian Theatre, was approached to take part, but declined.[40] Moore had been chief surveyor to the Dutch dredging expert Cornelius Vermeyden for a period of seven years in the 1650s. Their fen drainage scheme was a large and controversial one, spanning twenty years, under the administrations of two monarchs and a commonwealth. It involved the introduction of sluices (adjustable barriers to hold water levels), and the cutting of new straight channels to shorten rivers (including the 21-mile New Bedford river). At one point the workforce engaged in digging the channels and banks numbered up to 10,000 men. Investors received land once the drainage was completed: a total of 95,000 acres (approximately a quarter of the area of the Level) was divided up amongst them.[41]

Moore's background in applied mathematics stood him in good stead on the fen drainage project. His 1650 *Arithmetick* handbook, 'treating of vulgar arithmetick in all its parts', stressed throughout

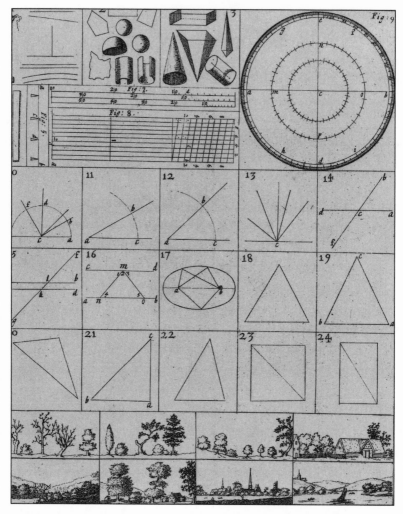

12. Surveying techniques from Moore's *New System of Mathematics*

the usefulness of its instruction for practical enterprises. Discussing the use of decimals with enthusiasm, for instance, Moore declares:

In practical Arithmeticke, if the first institution had been in Decimalls, we had never been troubled with so many fractions; it were yet worthy the name of Reformation to cause the

fractions of mony and weight to be altered: And as concerning
the ease in measure, Surveyors and land-meaters [sic], who use
the decimall chaine, and those who use a decimall foot yard or
scale can best certifie upon experience.[42]

The full title of Moore's *Mathematical Compendium*, compiled by his
student Nicholas Stephenson in 1674, and republished several
times, makes its vigorous applied nature yet more apparent:

> A Mathematical Compendium: Or Useful Practices in
> Arithmetick, Geometry, and Astronomy, Geography and
> Navigation, Embattelling, and Quartering of Armies,
> Fortification and Gunnery, Gauging and Dyaling.[43]

At the end of his life Moore was involved in producing an applied
mathematics textbook for the boys of Christ's Hospital School, an
institution founded by Moore, Pepys and others to provide instruction to prospective seamen, navigators and surveyors.[44]

For the fen drainage project Moore was required to survey the
terrain accurately before and after drainage, to calculate the opti-
mum positions for ditches and sluices, and to draw up the
boundaries of the parcels of land made available for cultivation by
the drainage. Aubrey claimed that Moore's understanding of the
movement of tidal waters helped him to devise a novel arrangement for the banks 'against the sea' in Norfolk 'for he made his
bankes against the sea of the same line that the sea makes on the
beach' – orthogonally to tidal flow, rather than in a straight line
along the beach.[45] Moore continued to be extremely proud of
the accuracy of his surveying in the Fens. When Charles II
approached the Royal Society in 1670 with a request that they
match the Paris Observatory's success in measuring an arc of the
meridian with a comparable English measurement, the first proposal made was to take advantage of one of Moore's man-made
features in the Fens. It was agreed that the measurement could be
made along 'Bedford-river, about twenty miles in length, formerly

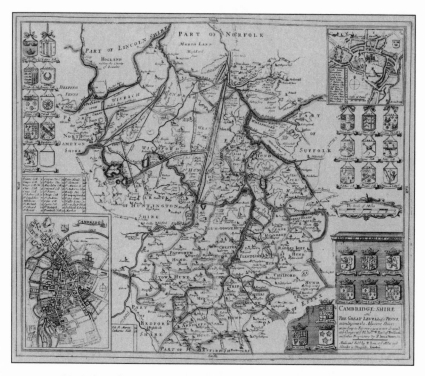

13. A sheet from Moore's *Mapp of the Great Levell of the Fens* showing
drainage canals

surveyed with exactness by Mr Moore'. (Because of lack of
funding the measurement was never in fact carried out.)[46]

Between June and September 1663, Moore took part in a preliminary survey of Tangier – a feasibility study for the proposed
fortifications. Pepys (closely involved from the Navy Office)
recorded in his diary in May of that year that the committee which
had been put in charge of the project were to consider 'sending
Captain Cuttance and the rest to Tanger to deliberate upon the
design of the Molle before they begin to work upon it' (Cuttance
was Sandwich's flag-captain and a committee member). On their
return, each member of this survey group was paid a hundred
pounds for their 'paynes in setting out the Bounds of the Mole &c'.

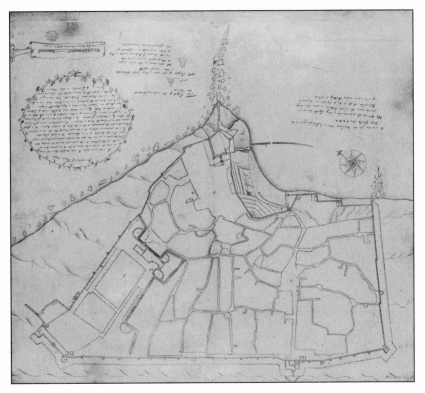

14. Jonas Moore's map of the Tangier fortifications

They presented the committee with 'a brave draught of the Molle to be built', adding that when completed it was 'likely to be the most considerable place the King of England hath in the world'. As with the fen drainage project, lucrative spin-offs were to be had by those closely involved. Moore's wealth in later life – which enabled him to act as generous benefactor to the Greenwich Observatory – undoubtedly came from the official and unofficial remuneration he received from his surveying work at home and abroad in the 1650s and 1660s.

The 'draught' or preliminary drawing of the proposed Mole, executed by Moore himself, was subsequently engraved by Wenceslaus Hollar, and published in 1664 for promotional purposes, as a kind of early investment brochure. In its printed form the map is more than four feet by two, made up of three sheets, and bears the title 'A Mapp of the Citty of Tanger; with the Straits of Gibraltar. Described by Jonas Moore, Surveyor to his Royall Highnes the Duke of York'. It also carries the royal arms and a dedication to the King and Queen by 'their obedient servant J. Moore by the Commaund & appointment of the Lords Commissioners for the affaires for Tanger'. Pepys liked it enough to frame it: his diary entry pronounced it 'very pleasant and I purpose to have it finely set out and hung up'.[47]

Over the period of the Tangier Mole's construction, a number of engineers, draughtsmen and engravers became involved speculatively in producing and marketing graphic versions of the fortifications and the surrounding landscape. In 1669 Wenceslaus Hollar

15. Wenceslaus Hollar's watercolour view of the Tangier Mole

petitioned successfully to accompany Henry Howard's expedition of that year as 'royal scenographer' (for which six months' work he was paid £100). On his return he published a series of etchings (*Divers Prospects in and about Tangier* [1673]), and three panoramic views. Sir Hans Sloane acquired six of Hollar's original drawings for his collection of travel documents and foreign curiosities.[48]

In 1675, Robert Thacker, who was employed by the Ordnance Office for 'draweing draughts of his Majesties Principall Fortifications' published his 'accurate map of Tangier and its Mole'.[49] A year later Moore employed Thacker on a project of his own, to produce a set of engravings of the new Greenwich Observatory. These fine engravings (after drawings by Francis Place) meticulously document both the building and its

instruments. Since Moore had personally paid for the instruments, and had been responsible for raising the money for the building, the images form a kind of public record and acknowledgement of Moore's achievement (as his earlier maps did of his fen drainage activities, and his Tangier Mole involvement). The prints were probably ready for sale at the time of the observatory's opening in summer 1676 – two months before, Hooke records in his diary that Moore had shown him a picture of 'the front of Greenwich Observatory'.[50] Pepys owned a complete set of twelve plates.

Large-scale wall-maps and charts of the kind the Dutch liked to hang on their walls did not catch on to the same extent in England. Nevertheless, Moore and his fellow-surveyors and draughtsmen evidently recognised a certain demand for these state-of-the-art graphic representations of innovative schemes in progress.

In 1666 it looked briefly as if Jonas Moore might take over responsibility for the Tangier fortifications. Work was progressing unacceptably slowly, and word began to be put about that the committee had come up with a replacement for Major Samuel Taylor, the officer in charge of construction (who, in any case, was threatening to resign). The leading on-site member of the team, Sir Hugh Cholmley, back in London briefly, reported to Taylor:

> Circumstance found out the man meant was Surveyor to his Royal highness Mr Jonas Moore by name who has bin sent in Tangers at first, to set out the Mole, one much esteemd by Sir William Coventry and of whome I have so good A Character that for ought I yet know, if upon your next Letters I finde your resolution does continue to be gon, I shall engage my selfe to Mr Moore and bring him over with me towards Christmas.[51]

Taylor, however, was given a salary increase (to £500 a year) and did not resign. By the end of the year, in any case, Moore had his hands full as a leading member of the team of surveyors dealing with the aftermath of the Great Fire of London. Moore acted in an advisory capacity on the Tangier fortifications for one final time in 1669, when he and Wren certified the arrangements proposed for supervising the further construction work as 'exact and Sufficient'.

It was Moore's involvement with the Tangier Mole construction plan that first brought him into contact with the Royal Society. Just before he set out on his first trip there in 1663, a colleague approached the Society for 'some leads and balls for sounding without a line for one Mr Jonas Moore going to Tangier'; these were duly supplied, and Hooke described the device more fully for the Society's benefit on 30 September. In December, Thomas Povey reported to the Society that 'Mr Jonas Moore had an engine for staying two or three hours under water in; and likewise a method of blowing up great rocks; both practised by him at Tangier'.[52]

In the end, expertise with explosives turned out to be of particular importance to the Tangier Mole project. In 1683, with 450 yards of the Mole completed, running east-north-east (but not the two hundred yards return, east-south-east), at a total cost of some £340,000, the project was abandoned. Rather than leave the fortifications to be used by Moroccan imperial forces, an expedition led by George Legge, Master-General of the Ordnance, and including Pepys, was dispatched to destroy the Mole and its fortifications. It is some sort of tribute to the skill with which the Mole had been constructed that it took until February 1684, and quantities of explosives, to complete the task of demolishing it.[53]

Impossible dreams and diving machines

Shortly after Jonas Moore's return from his first trip to Tangier, on 13 January 1664, he attended a meeting of the Royal Society with a request from one of the leading backers of the enterprise for assistance in developing a piece of technical equipment of vital importance to the Mole project:

> Mr Jonas Moore acquainted the council with Sir John Lawson's desire, that they would appoint a committee to examine Mr Greatorex's diving instrument, or to direct a good method of staying under water for a considerable time, to lay the foundation of the mole at Tangier.[54]

The instrument-maker Ralph Greatorex was already well known to the Society. He had been involved with the design of Boyle's air-pump at a preliminary stage, in 1658, although Hooke quickly dispensed with his services on the grounds that the machine he proposed was 'too gross to perform any great matter'. Still, however unsuitable for Boyle's experimental purposes, Greatorex's attempts at manufacturing airtight containers had evidently led him to construct some kind of diving-bell. Following Moore's announcement, Greatorex was called in and agreed to have his 'diving instrument' tested. He would provide a man to conduct the trial, as long as Lawson funded a new instrument and paid him handsomely if it was successful.

In May Sir Robert Moray reported that the trial had been unsuccessful, but the diver was ordered to keep practising. In June the diving-machine was tried again, in front of the Society, and the diver managed to breathe air successfully from a lead box for four minutes. We do not know if Greatorex's engine was subsequently tried at Tangier – Cholmley claimed that although Greatorex was 'a very ingenious person' he had a drink problem, which made him unsuitable for sensitive military work (though he did later obtain an Ordnance Office post as 'captain of miners and pioneers' in the third Dutch war, suggesting that by then he had moved on to even more dangerous work with explosives).[55]

The problem of enabling a man to spend long periods under water was one that the Tangier committee would badly have liked to

solve. It promised to make the laying of foundations for fortifications in deep water a considerably easier exercise. The obvious place
to go for the research and development to expedite production of
such an 'engine' was the Royal Society – which had already established itself as a kind of scientific information clearing house –
where at least they would know if any relevant experimental work
was going on in the field.

Boyle's air-pump work clearly had a bearing on the problems
divers experienced. In the early 1660s, however, his related air-pump
experiments, to monitor what happened to living things when the
amount of air in a closed container was decreased, were still at a
fairly rudimentary stage. Although tests had been carried out on
birds and small animals to show that these passed out when deprived
of air, but could (within limits) be revived when air was reintro-
duced, Boyle had not got far towards identifying the various
components of air (oxygen, nitrogen, carbon dioxide) and their differing functions in the process of breathing.[56]

Nor was Boyle's own work geared towards possible applications.
In the course of developing his thoughts on air-pressure, and why
we are not aware of the 'weight of air' on our bodies, Boyle had,
indeed, considered the analogous case of deep-sea diving. In his
'Relations about the bottom of the sea' he explained that he himself
'never pretended to be a diver': 'I do not pretend to have visited the
bottom of the sea; and none of the naturalists whose writing I have
yet met with, have been there any more than I.'[57] Relying on
recorded testimony of divers themselves, Boyle agreed, for once,
with Hobbes, that it was 'well enough known' that divers 'feel no
weight of water resting on them' – that they were (supposedly)
unaware of the weight of the water. He therefore set about accounting for this in theoretical works like his 1666 *Hydrostatical Paradoxes*,
by arguing that water, like air, exerted its pressure in all directions,
so that the diver did not sustain the full pressure in one direction,
thus remaining unaware of its effects.[58]

Given this kind of misinformation about the conditions under which diving was actually undertaken, it would be easy to imagine that the Royal Society's earnest armchair discussions of early diving-machine experiments conducted by Greatorex and others were simply impossible dreams, to be set alongside other unlikely pieces of information and evidence produced regularly at Society meetings.

In contrast to Royal Society members like Boyle, however, Edmond Halley did not have to ask the opinion of others when it came to matters to do with the sea. On 6 March 1689 he read a paper to the Society about a method for walking under water. He described how Oriental pearl divers were able to remain under water for several minutes, and how such divers had recently been employed in the West Indies to look for treasure in a Spanish wreck. On that occasion a diving-bell, Halley reported, had been used to refresh the divers so that they did not need to return to the surface, but they still had to hold their breath while they worked. Halley proposed making a bell that could be moved around on heavily weighted wheels under the water so that the men could work from inside it. He also proposed that additional air could be brought down to replenish the supply under the bell in strengthened casks.[59]

The reference to the Jamaican wreck was not mere hearsay. Halley's informant was Sir Hans Sloane, physician to Christopher Monck, Duke of Albemarle, who in 1689 had just returned from two years in Jamaica (where Albemarle had been appointed Governor in 1687).[60] Shortly after Albemarle's arrival to take up his post he was told about a 'great Plate-Wreck' which had recently been located – a Spanish treasure ship that had sunk in 1659 'to the North-East of Hispaniola'. Albemarle immediately set up a company, with a number of other investors, to finance an expedition to salvage the ship's cargo. The venture was a tremendous financial success. As Sloane described it later in his published account of his Jamaica voyage:

They found this Wreck, and wrought on it till the Ships Crew
grew scarce of Provisions, when they had taken up about
Twenty six Tuns of Silver. Sloops and *Divers* were sent, who
took up a vast quantity more of Plate and Money, so that
before a second Fleet came from *England,* the greatest part of
what Silver remain'd unfish'd was taken up.[61]

According to Sloane, Albemarle made a stupendous £50,000 on his
initial investment of £800. As a result, 'not only the first Patentees,
but many other People, hoping for the same Success, took out
Patents for Wrecks lying at the bottom of the Seas in all places'. No
wonder Halley was so keen on developing a workable diving-bell for
use on the sea bed.

Halley went on working on his diving-bells whenever he could
find the time. His chance to capitalise on his new technology
came just two years later. In April 1691 the *Guynie,* a frigate of the
Royal African Company, sank off the Sussex coast near Pagham.
On her previous journey from West Africa the *Guynie* had carried
a cargo of gold. When she made landfall at Falmouth on her next
trip, in February 1691 her captain, William Chantrell sent a message to the Admiralty requesting a man-of-war to convoy *Guynie*
along the south coast, suggesting that she once again carried
African gold (England was at war with France, making lone ships
vulnerable to sorties into the Channel from the French mainland).

The records of the Royal African Company reveal that *Guynie*
indeed carried a valuable cargo. A letter from three merchants on the
Cape Coast (the Gold Coast) notifies the London office that the
vessel carried a considerable quantity of gold, which was being
brought to England on behalf of several Portuguese merchants and
others, to be delivered to the Company. The merchants also
enclosed a bill of lading for '184 elephants' teeth [tusks] of twenty-
five hundred [weight] gross'.[62]

16. Sir Hans Sloane

A month after the request for an armed escort, Chantrell, at
Chichester, reported the loss of *Guynie* to the Company head office.
On 8 April the Company instructed that the ship and her cargo
should be salvaged, and ordered that an armed guard be mounted

to prevent looting. By 22 June 1691 Halley was at work with Chantrell, attempting to raise the gold and ivory. In September a company was set up and a patent granted to Halley and three other investors for 'the use of an engine for conveying air into a diving vessel, whereby they could maintain several persons at the same time to live and work safely at any depth for many hours together'.[63]

Halley's attempt to salvage *Guynie*'s cargo continued for the next five years. As far as we know he succeeded in raising only a single elephant tusk, on 21 March 1695. In 1694 Halley apparently formed a joint-stock company to raise additional money for his diving-bell and the wreck-salvage operations: a weekly paper of 20 July 1694 reviews the prospects of selected joint-stocks and singles out a diving stock as a good investment, adding that 'if Mr Halley should succeed, of which (were the wars at an end and the seas secure) he seems very sure, the profit would be very considerable'.[64] On at least one occasion Halley's salvage operation did suffer from the hazards of unsafe times. Pepys reports in his Naval Minutes 'Mr Halley's having his vessel taken from him by a privateer when he was at work in diving upon a wreck somewhere upon the coast'.[65]

In the course of Sir Hans Sloane's published account of Albemarle's highly profitable, comparable salvage operation in the West Indies, Sloane sheds light in passing on how it was that Halley, in spite of his remarkable equipment, was unable to rescue *Guynie*'s cargo and make his fortune. Some of the silver Albemarle's team raised in the West Indies, he reported, had already become irretrievably corroded, up to a quarter of an inch thick (Sloane kept some pieces-of-eight thus encrusted in his collection of curiosities). Yet more of a hindrance to salvage, he continued, was driven sand, which rapidly concealed the wreck. This was what had caused the failure of Halley's salvage attempt:

17. Halley's diving-bell in operation

I. Greenwich Observatory by Robert Thacker and Francis Place. Thacker was the Ordnance's sketcher of fortifications

2. Christiaan Huygens

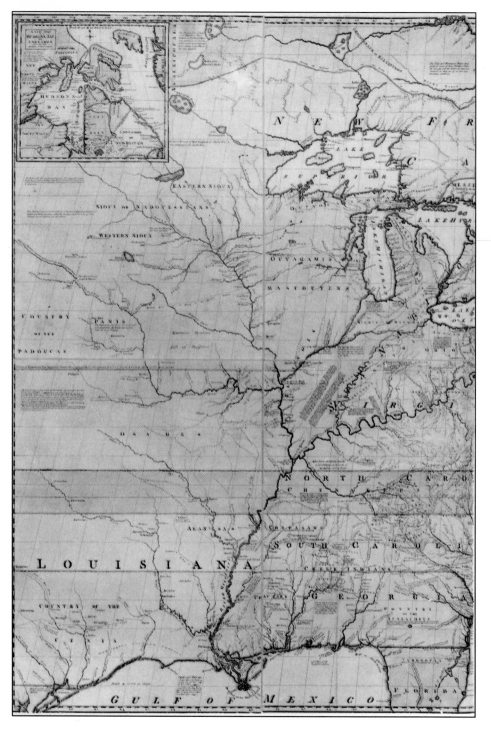

3. Mitchell's map of British and French territories in North America

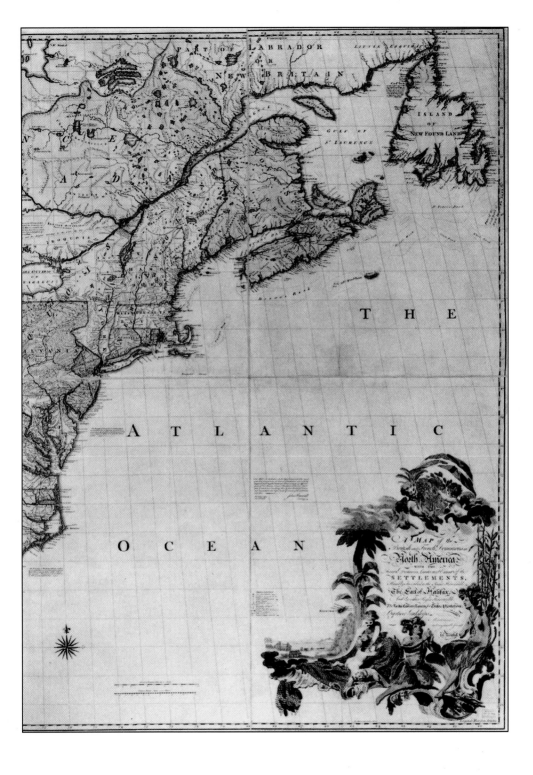

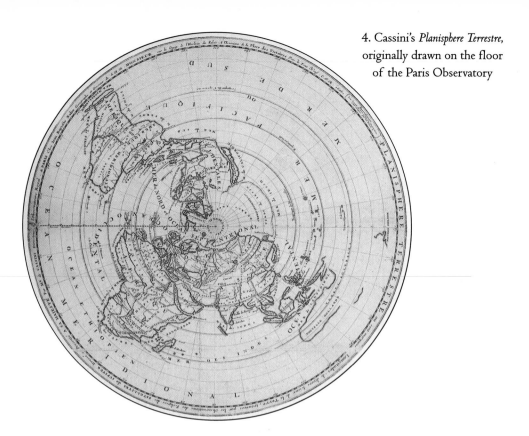

4. Cassini's *Planisphere Terrestre*, originally drawn on the floor of the Paris Observatory

5. Pressed specimens
of Cape plants from
Hermann's collection

BELOW
6. Wenceslaus
Hollar's watercolour
view of the Tangier
Mole

7. The Great Room at Greenwich, with clocks

8. Tompion clocks at Greenwich

9 & 10. Crab illustrations by Merian for Rumphius's *Indonesian Natural History*. Shell illustrations by Merian

11. Tulips and a gooseberry by Merian

12. Various insects by Merian

13. Two beetles by Merian

14. A Surinam lizard on a manioc plant, observed *in situ* by Merian

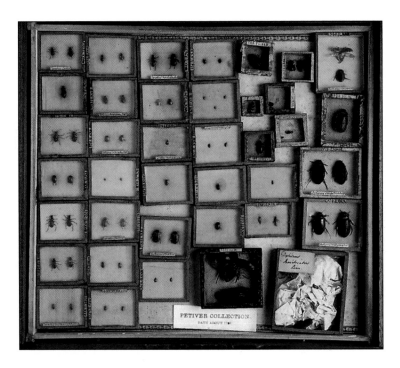

15. Original box of beetles from Petiver's collection acquired by Sloane

It is not only from corrosion, but also Sand driving by the
Winds and Currents, or Earthquakes that happen at the
bottom of the Sea, that Wrecks may be cover'd, and past
finding out. I remember an *African* ship, laden with Elephants
Teeth, wreck'd on the Coast of *Sussex*, which Mr. *Halley* told me
was in a very short time almost covered with Sand and Ooze,
so that the Project of recovering the Teeth, was frustrated,
though by the help of a Diving-Bell, contriv'd by his
extraordinary Skill, they had gone to the bottom of the Sea,
and into the Ship, where they had a perfect view of the ship,
and all about it.

As a result, Sloane concluded, those who had invested in *Guynie*'s salvage, taking Albemarle's success as their model, had lost large
amounts of money.

Nonetheless, Halley's five years' work on salvaging *Guynie* led to
modest, experimentally tested, scientific advances in diving-bell and
diving-suit construction (not least because the joint stock company
provided Halley with funds to develop his innovative technology),
and appears as such in the annals of the Royal Society. The salvage
operation led to significant improvements in Halley's diving-bell,
and the first diving-suit of a kind still used in deep-sea diving today.
On 26 August 1691 Halley read a paper to the Royal Society
describing the success of his diving experiments. He described his
bell, and how he replenished it with air. The bell was five feet in
diameter at the bottom (where it was open), three feet at the top,
with a thick glass window, and a small cock for letting out foul air,
and five feet deep. There was a bench for the divers to sit on, and he
reported that he kept three men under water at 18 metres for an
hour and three quarters.

As the bell was lowered, the air inside it was compressed, and the
water rose inside it. To counteract this, Halley brought air down to
the bell in strengthened casks, with iron bindings, weighting them
with lead so that they sank. The casks had a cock at the top and a

bung-hole at the bottom. Once inside the bell, the upper cock was opened, and the air was driven out of the cask, into the bell by the water pressure. He reported the effect of increased pressure on the divers' ears, and how the pain could be lessened by oil of sweet almonds.

In a second paper, delivered to the Royal Society on 23 September (the same day Halley read the paper on the transit of Venus, which later led Mason and Dixon to the Cape to observe it), Halley described his diving-suit, and claimed priority for inventing it. During October he enlarged on this first description, demonstrating the pipe he used to take air into the diver's weighted helmet, and the thick flannel suit the diver wore under his suit against the cold experienced in deep water. These papers make it clear that Halley had himself used the suit to dive off the Sussex coast.[66]

The range of the Royal Society's activities and types of interest is vividly illustrated by the contrast between Halley's diving-bell experiments and Boyle's discussions of the effects of altered air-pressure on respiration. Neither kind of activity can be singled out either as decisively a scientific breakthrough or as the obvious line of development for the future. But somehow out of the rich mix came the attitudes and expertises that produced the kind of science we recognise today.

CHAPTER 6

STRANGE SPECIMENS

The garden of creation

WHEN CASSINI AND Flamsteed encountered faraway places like Cayenne and the Cape, as they compiled their comprehensive astronomical tables, these were no more to them than names on a map. The same is true for the cerebral and little-travelled Newton. The tables on which Newton based his improved computations of comets' orbits, included in the second edition of his *Principia*, assembled material from observers at sites as far afield as Maryland and the Cape of Good Hope. For orbital calculation purposes, these locations were no more exotic than Manchester, Edinburgh and Paris.

But for those who undertook long and difficult journeys to make the observations, the encounter with distant lands and previously unknown cultures was a revelation, and (to judge from their diaries and memoirs) it remained a continuing source of amazement. The very temperament that made men like Richer and Halley energetic observers of the heavens also made them enthusiastic documenters of their novel surroundings. Scientists who travelled the trade routes to conduct barometer and pendulum experiments on the tops of distant mountains, or to check the sea-going reliability of new-fangled clocks, readily extended their scientific investigations to study the unfamiliar world around them. Theirs was a more attentive eye than that of the habitual traveller. They were, moreover, trained in precision. They recorded in notebooks the diversity of flora and fauna, of landscapes and rock formations, of dress and customs; they collected living specimens; they pressed plants, pickled animal parts, pinned butterflies and emptied shells.

An abundance of exotic and unfamiliar material flooded into northern European salons and scientific communities. It fuelled avid collecting of rarities and curiosities, by those with the means to

1. Exotic seventeenth-century Dutch still-life with tulips, insects and a lizard

do so, and opened up whole new markets in unexpected items like exotic fruits and menagerie animals. During the 1620s and 1630s, the Dutch turned the trade in tulip bulbs into a highly profitable, speculative commercial area, in which prices became hugely inflated, and importers manipulated the 'scarcity' of highly sought-after varieties by destroying stock. (The bottom suddenly dropped out of the tulip market in 1637.)[1]

But access to the exotic and strange in nature also led to more serious scientific projects. When Guy de Tachard and his fellow Jesuits made landfall at the Cape *en route* for China, in 1685, the accommodation with which they were provided by the Dutch East India Company was situated on the edge of the Company's thriving botanical garden:

> We were mightily surprised to find one of the loveliest and most curious Gardens that I ever saw. It contains the rarest Fruits to be found in the several parts of the World, which have been transported thither, where they are most carefully cultivated and lookt after.[2]

The travellers had good reason to be amazed. A well-ordered, assiduously cultivated and carefully tended garden, displaying rare and wonderful species from all over the world, was not what they might have expected at this most dangerous, and supposedly barren of spots on the navigational charts. Nor were they exaggerating the richness and diversity of the fruit and vegetables growing there. Their guide to the garden, Hendrik Adriaan van Reede tot Drakenstein, the Dutch East India Company's commissioner who had arrived on a mission to inspect the Colony only shortly before the Jesuits, described the garden with equal enthusiasm himself on his very first evening there (19 April 1685):

> Nothing can be compared with the pleasant aspect of the Company's garden, the mere sight of which is able to comfort and refresh a man coming from that wild, raw, and merciless

sea in a treeless land of bare highland. For the foot-paths, which extend further than the eye can see, are planted on either side with high walls of pleasant green trees, where one can be safe in the strongest south-easterly winds. The whole garden is divided into a great many square sectors or beds, planted with all sorts of fruit-trees and vegetables which, protected against evil winds, grow and flower plentifully, so that benefit and pleasure are here combined.[3]

Two days later, van Reede enlarged on his first impression in his official report. The garden, he recorded, had an area of about forty acres. The ground was divided into sixteen to eighteen sectors or fields by three broad walks, and eight or nine cross-paths:

These fields are planted with Cape shrubs or with alders or bay-trees, all trimmed to the same height and thickness, 20 to 22 feet high and 6 to 8 feet thick. Between these high and thick hedges the fields are enclosed. Several of them are planted with European fruit-trees, such as apples, pears, cherries, peaches, as well as oranges, lemons, and pomegranates in the form of orchards, also along the paths and broad walks, and they are very luxuriant and thriving. The others are sown and planted with all sorts of greens, vegetables and roots.[4]

According to van Reede, the garden was tended by fifty-four slaves (men and women), under the direction of a Dutch head gardener. All were housed together in a solid, square brick building, which also contained the apartments of the head gardener and drying-rooms for storing the seeds, roots and fruits. It was here that Tachard and his fellow-Jesuits were invited to set up their 'observatory'.

By the 1680s the Cape Colony's botanical garden was already one of the finest in the world. It dated from the foundation of the Colony by the Dutch East India Company in 1652. On 7 April of that year Jan van Riebeeck and Hendrick Hendricxen Boom landed at the most southerly point of the continent of Africa, with

2. Pressed specimens of Cape plants from Hermann's collection

instructions to establish a suitable staging-post for Company ships on their way to and from Indonesia and the west coast of India. The Company Gardens they founded (and of which Boom became the first master gardener) were intended to provide fresh produce for Company ships. Within a generation they were also serving as an experimental laboratory for exploring the medicinal properties of indigenous plants (to avoid the expense of shipping basic medical material all the way from Amsterdam).

At the time of van Reede's visit in 1685, the varieties of plant cultivated in the Cape Gardens were almost entirely imported – either from Europe, or from Dutch East India Company colonies further to the east in India and Ceylon. Little had been done in the way of systematic research on native plants and trees, although individual botanists had already begun compiling floras of the region for scientific purposes. (Paul Hermann compiled the first collection of dried and pressed Cape plants in 1672 – his 'dried garden' later came into the possession of Sloane, and thus became part of Sloane's extensive plant collection.)[5] A month after first planning the garden, in 1652, its creator, van Riebeeck, had established the precedent by writing home to Amsterdam requesting that he be sent a range of specimens they might attempt to grow for produce – bearing in mind that the object was for this produce to be carried conveniently aboard ship as provisions on long sea journeys:

> Seeds or plants of sweet potatoes, pine-apple, watermelon and
> other melons, pumpkins, calabash, cucumbers;
> India radish, and all other imaginable seeds;
> All kinds of young trees, shoots and sprouts.
> Of vines also which will thrive as well along the mountain
> slopes as in Spain or France.
> Oranges, lemons, pampelmooses, banana, mango, and
> mangosteen, durian, sour-sop, seri [lemon grass] and pinang
> [betel-nut palm].

Coconuts of which we had 12 to plant.
Bamboo plants or slips or seeds will be useful in many ways as
there is suitable forest here to make anything of.
Guavas, pomegranate, pawpaw.
Sugar cane or its seed.[6]

Van Riebeeck's 'shopping list' shows that the Cape gardeners
treated their site as a *tabula rasa* – a botanically entirely empty land on
which they would test plants that already flourished elsewhere. After
several experimental seasons, the market-garden character of the
garden was well established. Potentially money-making crops like
rice and coconuts failed to thrive in the Cape soil and climate.
Almonds, apricots, blackberries, cherries, chestnuts, figs, lemons,
mulberries, oranges, peaches, pears, plums, pomegranates, quinces,
strawberries and raspberries, on the other hand, all did extremely
well. No wonder Tachard was impressed; no wonder, also, that the
Jesuits were able to recognise this as, by and large, a familiar kind of
earthly paradise.

In the 1685 report he wrote to head office, van Reede advised
that the Cape's climate could be exploited fully commercially only if
more attention were given to the indigenous plants, which clearly did
better in the local climate than imported varieties: 'One very clearly
saw here the different nature of the plants and trees, for those from
Europe began to lose their leaves when the monsoon set in, while on
the other hand those from these regions sprouted young shoots
and leaves again and produced their blooms and flowers.' Van Reede,
who had spent more than ten years as the Dutch East India
Company's commander at Cochin on the Malabar coast of India,
counselled greater exploitation of the Cape's indigenous flora for
medicinal purposes:

This protruding corner of Africa is also notable with regard
to plants and trees, for wherever one goes, on low and moist,
high and dry lands, everywhere one finds great plenty of

3. Garden of earthly paradise, with tulips, pineapples and other exotica

fragrant flowers of the most brilliant and beautiful colours
that one might imagine, most of which have bulbs such as are
not found in any other countries of the world. So that
doubtless this soil brings forth many powerful herbs serving
man's health if they were known and used in the right way,
which will also be profitable for the Honourable Company to
be examined.

Finally, to complete his survey of commercial opportunities
based on the Cape's botanical strengths, van Reede suggested that
the Colony could corner the eastern market in wine, brandy and
vinegar by improving the quality of Cape wine, which he thought
would be less sour if the skins, pips and stems were removed from
the grapes before pressing them: 'for since we find that these regions
of Africa near the headland are as suitable for growing vines as any
countries in Europe, the Honourable Company might be able to
provide for the whole of the Indies from here'.

Van Reede was entitled to speak with authority on botanical matters. Between 1670 and 1674, while he was commander of Cochin
(on the west coast of India), he had been responsible for putting
together a monumental illustrated guide to the entire flora of Malabar,
the *Hortus Indicus Malabaricus*. This comprehensive survey in twelve volumes, published between 1678 and 1693 in Amsterdam, boasted no
fewer than 1,794 beautifully executed plates, and is generally regarded
as one of the greatest Floras of the seventeenth century.[7]

The idea for the Malabar botanical project had come out of advice
to van Reede from Andries Cleyer, the senior physician in charge of
medical supplies at the neighbouring Dutch East India Company
settlement at Batavia (Jakarta). Cleyer noted that important medicinal plants grew in quantity in Indonesia and Ceylon – in particular,
colocynth apples in the Jaffnapatnam region, and sarsaparilla near
Colombo. 'There are so many resources here in case of illness,'
another Company physician reported, 'that one could produce not
only a whole list of familiar medicinal plants, but moreover a whole

4. Frontispiece to Breyne's *Exotic Plants*, showing Cape plants

book of the unfamiliar ones.' Cleyer set up a local laboratory, where (in addition to conventional preparations using materials brought all the way from Amsterdam) he started to produce remedies using local ingredients, and supplying them to other Company settlements nearby – Ceylon and the Cape.

The suggestion that much-needed medicines might be manufactured locally, using local materials, to be prescribed for the many fevers and other illnesses from which Company personnel suffered seemed a good one. Commercially, too, the discovery of medicinal plants, growing in abundance, looked extremely attractive. As the Dutch East India Company extended its activities in Asia, it became clear that they needed to diversify beyond the spice trade to sustain their existing level of profit. The Company's monopoly of Malabar pepper (as well as important cinnamon and opium interests), in particular, was threatened around 1670 by unrest in Malabar. The Company added commercial production of medicinal raw materials to the other proposed extensions of their Asian operations, mining (particularly for copper) and plantations.

Van Reede quickly took a close personal interest in the flora of Malabar. He built a laboratory in his official Cochin residence, and set Paul Meysner to work with a furnace and distilling apparatus. He was instructed 'to extract from famous medicinal herbs, fruits and roots, their waters, oils and salts, and to examine in what respect they are equal to or excel the European ones, in order to provide the Company's medical shop therewith and to avoid the annual expense of many ineffective leaves, roots, seeds, and ointments from Batavia to Ceylon'.[8] Meysner was particularly successful in developing a process for distilling cinnamon oil from wild cinnamon.

Between 1674 and 1676 van Reede undertook, and personally financed, a comprehensive inventory of Malabar plants. By now a passionate botanical enthusiast himself, he used local labour to collect specimens, and consulted with teams of local experts to identify them, and to describe their medicinal and other properties. The plants, their descriptions and therapeutic value were then all recorded alongside faithful images of the plants, drawn from live specimens. The whole undertaking depended crucially on close collaboration between the European botanist and local experts in plants and medical materials. As van Reede later recalled:

> The plants were subjected, if this was worthwhile, to the examination of skilful physicians and botanists, whom I have convoked for that purpose, both from my staff and from among [local] people of some reputation. Naturally I took care that the pictures of the plants were shown to this board, which sometimes consisted of fifteen or sixteen scholars, and that by means of an interpreter they were asked whether they knew those plants and their names and curative values.[9]

When we try to account for the important impact in this period of Asian drugs and medical treatment on European medical practice, we should look closely at the images and accompanying texts in these extraordinary and compendious botanical works. Important contributions by Asian pharmaceutical and therapeutic experts were made accessible to European physicians via ambitious compilations like the *Hortus Indicus Malabaricus*. The necessary expertise was readily available to van Reede, associated with existing Company operations, and he made systematic use of it. On the other hand, finding artists to provide Western-style drawings of the plants discussed was more difficult. Van Reede had to recruit his illustrators wherever he found natural talent. Of the four exceptional illustrators he used to record his plants, one was a soldier in the Cochin garrison, Antoni Jacobsz Goetkint from Antwerp.

Van Reede returned to Amsterdam in 1678 with his *Hortus Indicus Malabaricus* in manuscript. Before he set out, however, he left a complete copy of his work with physician and fellow botanist Willem Ten Rhijne, who had 'added in Latin the curative values and uses of plants' for van Reede:

Before I left I took care that the pictures of the plants were
depicted anew and that the descriptions which had been made
up to that moment were copied. I left this copy behind with
the distinguished Doctor Ten Rhijne, in order that, if either
shipwreck or the necessity to throw things overboard upset our
voyage, he might again send another copy of the Hortus
Malabaricus. But it pleased Almighty God to restore us safe
and sound to our paternal home.[10]

Van Reede's caution later inspired Georgius Everardus Rumphius,
the so-called 'Pliny of the Indies', to do likewise when he sent
home his vast study of the flora and fauna of the Indonesian island
of Amboina, in sections, starting in 1692. The ship carrying the
first six books was sunk *en route* by the French, and the precious
manuscript went to the bottom of the ocean. Fortunately, the back-
up copy Rumphius had prudently had made (he had ruined his own
sight, and now relied on amanuenses) could be sent to replace it;
because of its size, though it could not all be published at once, and
some parts languished for years in the archives of the Dutch East
India Company. Publication of Rumphius's life's work was finally
completed, under the guidance and sponsorship of a consortium of
publishers, long after the death of its author, in the 1740s.[11]

So, at least, the official story goes. In actual fact, many of the
original illustrations turned out to be missing from the duplicate
copy of Rumphius's text. In 1702 the botanical illustrator Maria
Sybilla Merian was commissioned to prepare sixty new illustrations
to replace them. Merian was a fine botanical illustrator, who had
established an international reputation with a sequence of volumes
of drawings of plants and insects, published in the 1670s, 1680s
and 1690s. At the time of the Rumphius commission she had
recently returned from a two-year self-funded trip of her own to
Surinam, accompanied by her daughter Dorothea, where they had
studied, collected and drawn the local insects and flowers. She was
desperately short of money, and no doubt delighted to take on

'ghosting' someone else's lost illustrations. Her first-hand familiarity with overseas exotica was probably one of the reasons she was recommended for the job of redoing Rumphius's drawings.

Merian combed existing Amsterdam collections for specimens that resembled the items Rumphius described, assisted by the editor in charge of the Rumphius volume, the Amsterdam collector Simon Schijnvoet. The resulting images were an artistic triumph, but somewhat less of a success in scientific terms. In a substitute plate containing three drawings of crabs, for example, one is a common Dutch shore species, another a rare Indian Ocean crab, drawn from a prized specimen in Schijnvoet's own collection — but not the one Rumphius described in his text. This did not, apparently, bother the publishers. The first Rumphius volume was published in Amsterdam in 1705 without any reference to the substituted plates, and with no mention of Merian's participation. In the same year Merian published a volume of drawings of Surinam specimens under her own name, subsidising it with her earnings from the Rumphius drawings.[12]

Back in Jakarta, Cleyer, the man who had kindled van Reede's interest in plants, moved on from pharmaceuticals to the lucrative business of garden rarities. In 1677 George Meister, previously gardener to the Duke of Saxony, arrived in Java. Cleyer and Meister worked together identifying unusual plants and fruit trees suitable to be introduced into Europe. The two men travelled to Japan together, where they collected a wide range of specimens that were 'acclimatised' in the Cape botanical garden. These included tea, camphor, guava, banana and pineapple. When Meister returned to the Dutch Republic he took with him seventeen chests of soil and plants, including three to present to William of Orange, and nine chests containing trees for the Amsterdam Botanical Garden.[13]

So when van Reede came to inspect the Dutch East India Company's Cape Colony in 1685, it was to be expected that he would take a serious interest in matters botanical. He found that the Cape Governor, van der Stel, who was something of an enthusiast

5. Crab illustrations by Merian for Rumphius's *Indonesian Natural History*.
One of these is in fact a common Dutch shore species

6. Shell illustrations by Merian

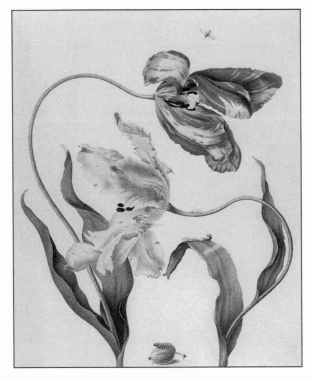

7. Tulips and a gooseberry by Merian

himself, had undertaken a project comparable with his *Hortus Indicus Malabaricus*. A complete Flora of the Cape region was being compiled, exquisitely illustrated by Hendrik Claudius, originally from Silesia, who had been dispatched from Jakarta by Cleyer to make an illustrated compilation of medicinally useful plants, and been retained by van der Stel to assist in his own project. Van Reede was captivated by Claudius's outstanding drawings. Tachard tells us that van Reede could hardly bear to let them out of his sight:

> We made acquaintance with a young Physician of *Breslon* in *Silesia*, called Mr. *Claudius*, whom the Dutch for his great Capacity entertain at the Cape. Seeing he hath already travelled into *China* and *Japan*, where it was his custom to observe every

8. Various insects by Merian

9. Two beetles by Merian

thing, and that he designs and paints Animals and plants perfectly well, the *Hollanders* have stop'd him there to assist them in making their new discoveries of Countries, and labour about the natural History of *Africa*.

He hath compleated two great Volumes in Folio of several Plants, which are drawn to the life, and he hath made a Collection of all the kinds which he hath pasted to the Leaves of another Volume. Without doubt the *Heer Van Rheeden* [van Reede] who had always these Books by him at home, and who shewed them to us, has a design to give the publick shortly an *Hortus Africus* after his *Hortus Malabaricus*.[14]

10. A Surinam lizard on a manioc plant, observed *in situ* by Merian

From Tachard's account it is clear that he and his fellow Jesuits spent a good deal of time inspecting Claudius's drawings, together with van Reede. And, indeed, at least two illustrations in Tachard's published account of his voyage can be identified as derived from Claudius's illustrations.

As it turned out, this encounter had unfortunate consequences for Claudius. Tachard had been correct in believing that, given the current political situation at the time of his visit to the Cape, the Protestant officials of the Dutch East India Company should have kept the Catholic missionary representatives of an enemy power at arm's length. Instead, their shared interests in botany and natural history drew them together into an enthusiastic and productive naturalists' collaboration. In any case, van Reede had had a great deal of help with his own work in Malabar from a Franciscan monk who had spent long years in Asia, and clearly felt that scientific research took precedence over religious or political affiliation.

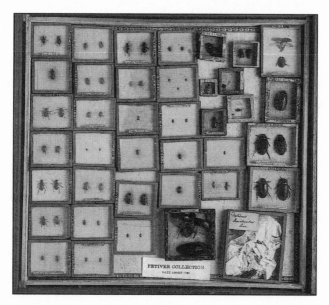

11. Original box of beetles from Petiver's collection acquired by Sloane

At the time of this cordial exchange of scientific ideas and expertise, however, Claudius had just returned from an official assignment, accompanying van der Stel on an expedition into the African interior reconnoitring for copper and other mineral deposits and raw materials for exploitation by the Company. There Claudius had acted as field illustrator, providing important pictorial material for the Dutch East India Company's research and development records. When Tachard's 1686 volume describing his voyage appeared in print, it became clear how much commercially sensitive, 'classified' information the French Jesuit had been given access to during his visit – much to the embarrassment of the officials at the Cape Colony.

To the authorities, the exchange of ideas and material – so vital for the growth of the sciences – was a matter of giving dangerous 'intelligence' to commercial and political rivals, and constituted deliberate or inadvertent industrial espionage. In June 1687, Tachard duly returned to the Cape, bringing with him an enlarged group of missionary-explorers, which included a botany expert, in the full expectation that he would be allowed to capitalise further on the scientific friendships he had formed two years earlier. He was bitterly disappointed. This time van der Stel made sure that he received just as much co-operation from Company employees as would preserve political decorum. When Tachard had left, van der Stel reported to Amsterdam, with some satisfaction:

> The French were evidently also very much disappointed that
> certain persons, suspected by the Governor of too much
> familiarity with them on the last visit, had in good time been
> sent away to Mauritius and Batavia. One was Pierre Couchet, a
> Frenchman from Amiens, who had managed to obtain a
> position as a gardener here. The other was Hendrik Claudius,
> an apothecary in your service. The understanding he had with
> the Jesuits has been fully shown, to our great perturbation, by

12. Cape of Good Hope chameleon

their book recently published regarding their Siamese voyage. In it his name is mentioned, and it is plainly stated that he communicated to them everything about the Colony and our inland expeditions, and perhaps more than we know of.[15]

Today, rather than applauding the restrictions van der Stel successfully imposed on the passing of information, we are more likely to admire the sheer stamina of travellers like Tachard, who were prepared to spend many months at sea, in serious discomfort and danger, in order to gain access to the latest scientific intelligence in Asia. Between 1685 and 1688 Tachard made two round trips to Siam, each taking over a year; most of that time will have been spent on the high seas, on the debilitating journeys there and back.

Almost the first thing Hans Sloane (not yet knighted) recorded in his naturalist's notebooks, on the voyage he made to Jamaica in the service of the Duke of Albemarle in 1687, was the morale-sapping

physical effects of seasickness. The ship set sail on 12 September;
it was almost immediately driven back to port by high seas and contrary winds. On 19 September Sloane wrote:

> We who had not been accustomed to the Sea, at first setting
> Sail, or even going aboard, by the Heaving and Setting, as
> Seamen term it, or the Motion of the Ship by the Waves, were
> almost all of us Sea-Sick. This first appears as a great
> uneasiness and load about the Stomach, disorder and aching in
> the Head . . . the Persons generally, from desponding and not
> caring what happens to them.[16]

No wonder that scientists like Huygens, who wanted to claim that
sea trials on potential longitude-measuring clocks had not been
conducted correctly, could convincingly accuse the operator of 'not
caring' on account of sea-sickness. Sloane, who had taken on an
additional paid job as doctor to all those on board during the trip,
went on to explain that his medical advice to sufferers was 'to drink
quantities of warm water, or Small-Beer', to act as an emetic, 'on
which they found Relief and in some days they grew better' (a
remedy still used by trawler-men, who feed quantities of porridge to
new recruits until they find their sea-legs).

As Tachard's encounter with Claudius at the Cape shows, naturalists and illustrators took every opportunity available to them to
exchange material for their mutual scientific benefit. Ten Rhijne,
who had assisted van Reede in his *Hortus Malabaricus* (and eventually
saw Rumphius's *Herbarium Amboinense* through the press in
1741–55), corresponded regularly from Java and Japan with Jacob
Breyne in Gdansk – an avid botanist, who never left Europe but
compiled well-annotated and exact botanical records of Asian flora.
When Breyne's Flora of exotic plants appeared in 1678, it included
a large number of illustrations and descriptions of Cape specimens,
derived from van Reede and Claudius, and supplied by Ten Rhijne.
(An eleven-page appendix by Ten Rhijne on 'De Frutice Thee' in

AD CAP. IX.

FRUTEX ÆTHIOPICUS
CONIFER, FOLIIS CNEO-
RI, SALICI ÆMULUS.

13. Early engraving of a distinctive Cape shrub

Breyne's work introduced readers to tea-preparation in Japan and its therapeutic properties.)[17]

As specimens and their images circulated in this way between travellers and those who remained at home, the boundaries between 'strange' and 'familiar' flora and fauna quickly disappeared. As far as the book-buying public were concerned, it was irrelevant that Breyne had never left Europe, as long as his *Cape Flora* contained pictures and descriptions of the surprisingly large number of plants found only there. Maria Sybilla Merian had travelled half-way across the world to observe Surinam flowers and insects in their natural habitat; her published Flora excited no more special interest than those compiled locally from travellers' specimens and other people's drawings.

Specimens were spread world-wide by naturalists, as freely and copiously as their watercolour and engraved images. During its early years, the Amsterdam Botanical Garden (established as a medical-materials resource by the city council in 1682) was particularly generously supplied with newly introduced species by the Dutch East India Company, who also helped finance the garden, which served as a shop-window and research institute for their imported pharmaceuticals. As a consequence of the Company's interest, the remarkable garden contained well over two hundred species indigenous to the Cape, compared with a mere thirty from the West Indies, and ten from North America.[18] The garden contributed to the training of surgeons and apothecaries, and plants could be purchased for the preparation of remedies. Lectures on the therapeutic properties of the plants were offered by the garden's own Professor of Botany – the first of whom, appointed by the burgomeisters of Amsterdam in 1685, was Frederik Ruysch (the same Ruysch whose collection of preserved, wax-injected, lifelike bodies and body parts attracted a steady stream of international visitors).[19]

Meanwhile, the Company shipped exotic bulbs (and even growing plants and trees) back in bulk from Asia, to satisfy the growing market for unusual garden and hothouse plants, rapidly turning the

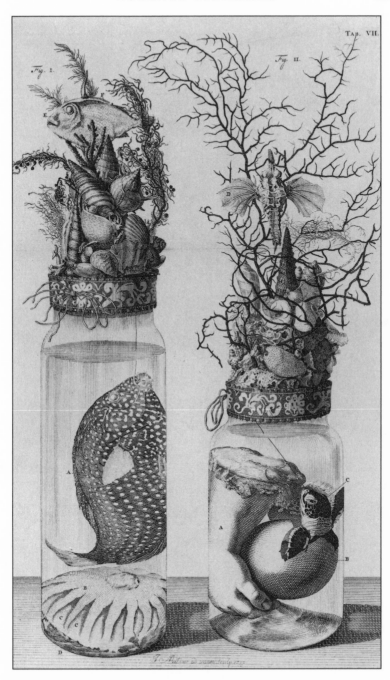

14. Ruysch specimens in glass jars

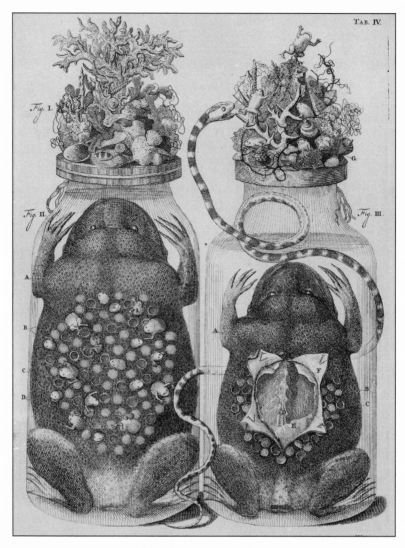

15. Ruysch specimens in glass jars

Dutch Republic into the international centre for market gardening it remains today. The fashion for exotica permanently altered the European landscape: the cultivated gardens of the wealthy henceforth boasted dazzling displays of tulips, narcissi, hyacinths, etc., as well as trees and shrubs never previously seen outside Asia. Around

1655, a Dutch East India Company ship returning from Japan, via the Cape, was wrecked off the shore of Guernsey. The ship's cargo was bulbs for the European market, among them a Cape lily – the so-called 'Guernsey Lily'. Bulbs of this lily were presented by the grateful crew to the Dean of Guernsey, after local people had rescued them. Today, the lily is regarded as a Guernsey native.[20]

Consumed with curiosity

Knowledge of these remarkable new floras and faunas was initially brought back to Europe, as in the case of Tachard, by the very same scientists who were conducting the astronomical experiments essential for research and development in mapping and navigation. At the French Académie, world botany became the field for which Colbert authorised the most significant funding after cartography. After Cassini and Huygens, the botanist Claude Perrault received the highest annual pension from the Académie, while considerable sums were given to finance the programme of plant engraving by Nicolas Robert and Louis Claude Chastillon, which was supposed to culminate in the Académie's major botanical project – a comprehensive Natural History of Plants.[21]

It was not simply the scientific community, however, that responded with enthusiasm to such a striking influx of botanical and natural historical material into Europe. Those with the wealth, or the prestige, to indulge a personal taste for the exotic and rare also took advantage of the novelties brought to their attention by the roaming naturalists. Louis XIV took a keen personal interest both in his extensive gardens, where fruiting orange trees grew in his greenhouses, and where 'eighteen Millions of Tulips, and other Bulbous Flowers' were planted over a four-year period in the gardens of the Royal Observatory alone. John Locke saw ostriches at Versailles, and an elephant that ate 'fifty lb. of bread per diem &

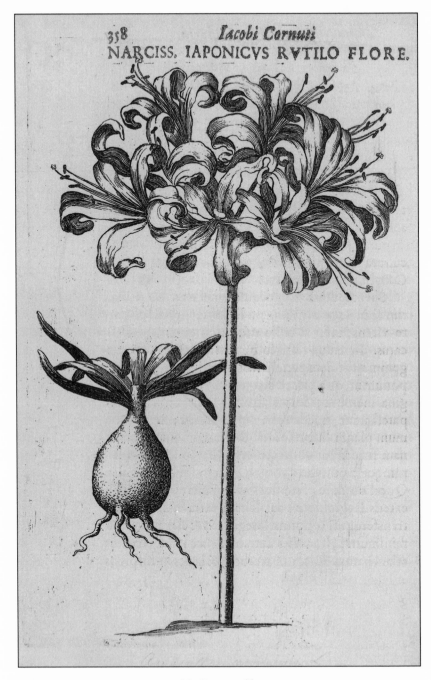

16. Guernsey lily

sixteen lbs. of wine and rice'.[22] (The menagerie at Versailles provided a steady, if unpredictable, supply of exotic animals for
dissection by the medical members of the Académie – the academician Claude Perrault died of an infection he contracted when he
cut his finger while dissecting a camel.)[23]

In September 1675 William III of Orange – a keen naturalist –
indicated to the Dutch East India Company in Amsterdam that he
would be delighted to receive 'all sorts of animals, birds, tissues, cabinets, and other curiosities' as an annual tribute. The Company
took the hint, recognising that the Stadholder's pleasure would help
oil the wheels of import regulation and Customs and Excise duties.
At the end of 1676 van Reede was instructed to collect birds,
plants, bulbs and seeds for William's 'cabinet of curiosities' in
Malabar. Rijklof van Goens Jr, Governor of Ceylon, received a
similar request from the Company to send a couple of living elephants, birds and other tame animals, cinnamon trees and pepper
trees, and rare plants to the Netherlands. In 1679 van Goens did
send elephants to Amsterdam.[24]

Glass-houses with built-in stoves for heating allowed resourceful
gardeners to grow tropical fruits, introducing Europe to tastes never
dreamed of in the days of mere apples and pears. Evelyn got his first
sight of a pineapple on 9 August 1661, recording in his diary: 'I first
saw the famous Queen-pine [pineapple] brought from Barbados presented to his Majestie.'[25] Pineapples were probably first artificially
produced in quantity in the hothouses of Sir Matthew Decker's
famous garden in Richmond. Decker was a well-known London merchant of Dutch origin, and president of the English East India
Company; he apparently enjoyed a 'truly Dutch passion for gardening'.[26] Contemporary books on horticulture suggest that a forty-foot
stove would ripen a hundred pineapples. The enormous cost of raising these hothouse pineapples was amply justified by the price that the
ripe fruit, with their unique texture and taste, could command. John
Locke, a keen gardener himself, used the pineapple as an example of

the impossibility of conveying direct sensory experience in words, in his *Essay Concerning Human Understanding*: 'He that thinks otherwise, let him try if any words can give him a taste of a pine apple, and make him have the true idea of the relish of that celebrated delicious fruit.'[27]

The rich of northern Europe were prepared to spend money on raising a pineapple from seed, or getting an orange tree to fruit, to astonish their dinner guests, on a scale the scientific community could not match. Nevertheless, their enthusiasm made a significant contribution to the growth in understanding of plants and plant growth. The great European gardens for the leisured of the seventeenth century, impeccably planned and maintained at enormous cost, sit alongside contemporary serious scientific research into therapeutic remedies, and botanical cultivation of Asian plants as raw materials for pharmaceuticals. Francis Bacon, in his essay *Of Gardens*, presents the garden in which no expense has been spared as the ultimate source of that repose and restoration of spirits which 'delight' can offer:

> And because the breath of flowers is far sweeter in the air
> (where it comes and goes like the warbling of music) than in
> the hand, therefore nothing is more fit for that delight, than to
> know what be the flowers and plants that do best perfume the
> air. Roses, damask and red, are fast flowers of their smells; so
> that you may walk by a whole row of them, and find nothing
> of their sweetness; yea though it be in a morning's dew. Bays
> likewise yield no smell as they grow. Rosemary little; nor sweet
> marjoram. But those which perfume the air most delightfully,
> not passed by as the rest, but being trodden upon and crushed,
> are three; that is, burnet, wild-thyme, and watermints.
> Therefore you are to set whole alleys of them, to have the
> pleasure when you walk or tread.

For Francis Bacon, the 'true pleasure' of a gentleman's garden combined expertise in assembling a rich diversity of growing things, from pineapples to daisies, with repose and delight.[28]

From the middle of the seventeenth century, visiting the gardens and cabinets of curiosities of the wealthy became a recreational activity in its own right for those with the means to move around Europe. John Evelyn, a committed English royalist, found himself with time on his hands during the Commonwealth years, and became something of an expert on other people's collections. He travelled extensively in Italy, and recorded his impressions of a whole range of gardens and collections of exotica in his diary. Back in England, on 17 September 1657, he visited the combined botanical garden and 'cabinet of curiosities' of the Tradescants, at Lambeth – popularly known, because of the extraordinary range and variety of its specimens, as the 'Ark'.

John Tradescant senior was a gardener by training who, during the 1620s and 1630s, had worked for some of the most powerful men in England, including Sir Robert Cecil and George Villiers, Duke of Buckingham, as well as both James I and Charles I. He designed elaborate and unique gardens for his employers, specialising in unusual plants and trees, some of which he introduced into England himself – he is credited with introducing (or possibly reintroducing) the horse-chestnut, which he used as an avenue tree for large estates when he found that the tree's fruit was not edible (contrary to what he had been told). The 'Tradescant cherry' appears in recipes of the period as an alternative to the morello cherry. John Tradescant senior made several plant-collecting journeys, in search of flowers and shrubs of distinctive beauty that could be successfully cultivated in England, travelling as far afield as Russia.

Under Buckingham's patronage he began to accumulate 'curiosities' as well as plant specimens, and to assemble a very considerable collection of rarities from all over the world, extending from coins and medals to ethnographic items such as Amerindian clothing and feather headdresses.[29] As time went on, Tradescant's trawling for rarities became increasingly ambitious, and he broadened the reach of his requests for donations of novel specimens by travellers of all

17. Tradescant's amber plum

kinds. In a letter addressed to Edward Nicholas, secretary to the Navy, Tradescant sent out a call to 'All Marchants from All Places But Espetially the Virgine and Bermewde & Newfownd Land Men' and especially 'from Gine or Binne or Senego Turkye Espetially to Sir Thomas Rowe Who is Leger At Constantinoble Also to Captain

Northe to the New Plantation towards the Amasonians' requesting
that they bring back rarities of all kinds for him: 'All maner of
Beasts & fowells and Birds Alyve or If Not Withe Heads Horns
Beaks Clawes Skins Fethers Slipes or Seeds Plants Trees or Shrubs,
& Also from the East Indes Withe Shells Stones Bones Egge-shells
Withe What Cannot Come Alive'.

After Buckingham's assassination in 1628, Tradescant bought a
house in Lambeth, established his own garden of botanical rarities,
and made both the garden and his extensive collection of curiosities
available to the public. By 1634 the Ark's accumulation of exhibits
had already reached such proportions that a superficial examination
of its contents took a keen visitor a whole day. To judge from one
such visitor's description, viewing the collection was almost as good
as travelling oneself: he was 'almost persuaded a Man might in one
daye behold and collecte into one place more Curiosities than hee
should see if hee spent all his life in Travell'. When Tradescant died
in 1638, his son, John Tradescant junior, who had also trained as a
gardener, followed his father's example of combining practical horticulture and naturalist specimen-collecting, making three trips to
Virginia 'to gather all rarities of flowers, plants, shells, etc.' to aug-
ment the collection further.[30]

The Tradescant Ark was accessible to all for the payment of a
small sum at the door. Among the well connected who visited,
Evelyn was not the only one to be entranced by the extraordinary
range of the Tradescant collection. In 1652 the younger John
Tradescant had employed the services of Elias Ashmole, a noted
antiquary and collector of coins and medals, to make a complete
inventory of the collection. Ashmole and the physician Thomas
Wharton itemised it systematically, and described its holdings in
suitably erudite Latin. A printed version of the catalogue was published in 1656 – a sufficiently unusual step to establish a mere
commoner's collection, the Tradescant Ark, as an international landmark, a must to be visited by all educated visitors to England, as

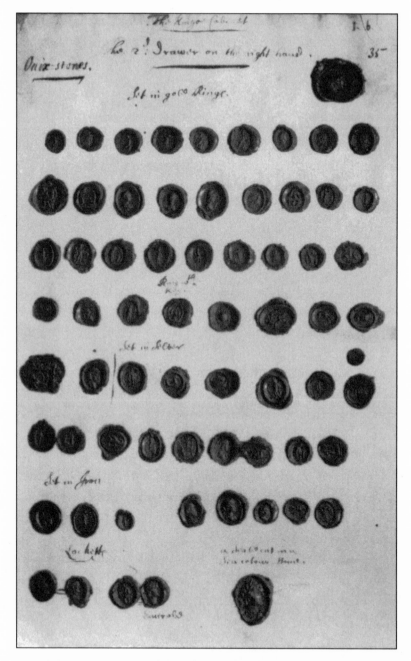

18. Wax impressions made by Ashmole of engraved gems in Charles II's
collection

well as a very considerable resource for botany and the natural sciences. At some point during the period of months it must have taken Wharton and Ashmole to examine and catalogue the Tradescant plants, coins and rarities, Ashmole developed an intense desire to acquire the entire collection himself. He negotiated a legally binding deed of gift with Tradescant junior, which stipulated that possession of the Ark would pass to Ashmole after Tradescant's death, perhaps in lieu of payment for the work Ashmole had done on the catalogue.[31]

Elias Ashmole was a London lawyer (the legal document he drew up with Tradescant turned out to be absolutely watertight), but by temperament a compulsive organiser, counter and classifier – a born taxonomist. He regarded himself as an expert numismatist (coin specialist), and at the end of his life had assembled a vast coin and medal collection of his own, put together over a period of more than thirty years, beginning in the 1640s (this was largely lost – nine thousand brass, copper and silver coins in total – in the great fire at Middle Temple in 1679). Ashmole was also a devoted supporter of the royalist cause, which is why he, like Evelyn, was without office, and had time on his hands in the 1650s. After he had completed the catalogue of the Tradescant collection, he was invited to catalogue the Bodleian Library's extensive collection of Roman coins – a task he later maintained was the most significant cataloguing accomplishment of his career. Immediately following the restoration of Charles II, in 1660, he was employed to prepare a complete catalogue of the King's coins, medals and gems.[32]

Even before John Tradescant junior died in 1662, he and his wife Hester were trying to overturn the deed of gift. After his death, relations between Ashmole and Tradescant's widow rapidly deteriorated, leading to a series of lawsuits by both parties. By this time Ashmole had bought and moved into the house next door to the Ark, and at one point he lodged a suit complaining that Hester's compost heap, piled too high against the wall dividing their properties, had allowed thieves to break into his own home. She in her turn complained that he had made a door in that wall without permission so that he could come and go as he pleased. On 18 May 1664, the Court of Chancery overturned a suit lodged by Hester, in which she attempted to regain control of the collections. The Court ruled that Ashmole should 'have and enjoy all and singular the said Bookes, Coynes, Medalls, Stones, Pictures, Mechanicks and Antiquities belonging to the collection', while Hester was to keep them in trust only during her lifetime, as provided in the original deed.

At the root of the dispute lay, in all probability, the problem of subsistence for Hester Tradescant after her husband's death. A manuscript catalogue of the 'curiosities' of another notable collector, William Courten, lists purchases made from Hester Tradescant, out of the collection, in 1667, at a time when she was prohibited by law from disposing of any of the contents.[33] We may assume that need, rather than malice, prompted such sales. On Ashmole's side, the move to the adjacent property in Lambeth and construction, during Hester's lifetime, of a means of access to the collection over which he had established legal control surely constitute harassment. By 1677 he had removed the most significant of the rarities to his own premises (he seems never to have taken any particular interest in the garden of botanical rarities, which had formed the original basis for the collection). On 28 April 1677, Hooke recorded in his diary that he had visited Ashmole's house, where he found William Dugdale (whose daughter was Ashmole's third wife). They 'saw Tradeskants raritys in garret. Saw Dees and Kethes and many other Books and manuscripts about, chymistry, conjurations, magick, &c.' Ashmole 'made [Hooke] exceeding welcome'.[34]

On 4 April 1678 Hester Tradescant was found drowned in the shallow pond in the grounds of her Lambeth home. Though the incident was buried as rapidly as the unfortunate woman's body, her death was widely assumed to have been suicide rather than accident. By July, Ashmole was the proud, official owner of the Tradescant

19. Ashmole's catalogue of the peers of England, reconstructed for
Charles II

Rarities, and had incorporated them into his own, far less distinguished, collection of books and curiosities. On 23 July 1678, John
Evelyn went again to Lambeth, this time to the house next door to
the Ark, 'to see Mr. Elias Ashmole's library and curiosities'. He
reported with some amusement that Ashmole was no intellectual
('but very industrious') and that his main scholarly interests seemed
to be astrology and the supernatural (studies that had no better reputation then than they do today): 'He has divers manuscripts, but
most of them astrological, to which study he is addicted, though I
believe not learned. He showed me a toad included in amber.'[35]

Ashmole's entire career was spent itemising, ordering and cataloguing the residue of the past, from the inventories of ancient
coins and medals, to his compendious *Institution of the most Noble
Order of the Garter* (1672), and the work he did in his official capacity as Windsor Herald authenticating, classifying and ordering by
rank the coats of arms of the English nobility. He cultivated the
great and the good, securing a good number of desirable additions
to his collections in the process. In 1674 Henry Howard, Duke of
Norfolk, gave him the gold 'lesser George' from his grandfather
Thomas Howard, Earl of Arundel's Garter insignia, perhaps in
recognition of Ashmole's work on the Order of the Garter.
Ashmole gave Howard pieces of the True Cross out of the
Tradescant collection – now his – in exchange (the lesser George is
still in the Ashmolean).[36]

Everything we know about Ashmole's personality and interests
should therefore prepare us for the next chapter in this somewhat
shameful story. Ashmole set out to use the extraordinary collection
of curiosities he had acquired to provide himself with a permanent
historical monument. Rumours that he intended to give his collection to the University of Oxford began to circulate around 1670:
Evelyn told his Somerset correspondent, John Beale, that the
academics of Oxford 'talke already of founding a Laboratorie, &
have beg'd Relighes of old Tradescant, to furnish a Repositary'. In

20. Elias Ashmole in full regalia

1675, the University accepted the rarities, and agreed to the condition, stipulated by Ashmole, that a proper building should be erected to house them. Thus was the great Ashmolean Museum born, and thus was the name of Ashmole, instead of Tradescant, preserved for posterity.[37]

It will come as no surprise to find that the preamble to the regulations for the Ashmolean Museum as completed (1686) makes no mention whatsoever of the John Tradescants, father and son:

> Because the knowledge of Nature is very necessarie to humaine life, health, & the conveniences thereof, & because that knowledge cannot be soe well & usefully attain'd, except the history of Nature be knowne & considered; and to this [end], is requisite the inspection of Particulars, especially those as are extraordinary in their Fabrick, or useful in Medicine, or applyed to Manufacture or Trade: I Elias Ashmole, out of my affection to this sort of Learning, wherein my self have taken & still doe take the greatest delight; for which cause alsoe, I have amass'd together great variety of natural Concretes & Bodies, & bestowed them on the University of Oxford, wherein my selfe have been a Student, & of which I have the honor to be a Member.[38]

Whatever, in the end, we think of Ashmole's motives, and the way he came by 'his' collection, the original contents of the Ashmolean Museum cannot, by any stretch of the imagination, be considered either systematic, or based on coherent selection criteria. This was an expert gardener's accumulation of 'other materials' garnered while pursuing horticulture for the immensely wealthy as the main project (ironically, Oxford already had a botanical garden at the time Ashmole proffered his 'gift', so the Tradescant rare plants were never considered as part of it). Ashmole's own considerable collection of coins and medals had been virtually destroyed by the 1679 fire. Nor was this *ad hoc* quality altered by the addition of other early bequests, by William Dugdale and Martin Lister, which were simply added haphazardly, to enlarge the holdings of England's first public museum (the Ashmolean continued the Tradescants' policy of allowing paying visitors, rather than, as elsewhere in Europe, restricting access to a hand-picked élite). As a collection, the Ashmolean said more about gentlemen's taste in curiosities than about the state of English natural science.

Putting things in order

Fortunately, the enthusiasm to acquire specimens on the part of gentleman collectors led some of the more scientifically serious and committed of them to be prepared to make strenuous efforts of their own to build up remarkable collections. The London physician Hans Sloane was already an amateur collector of pressed botanical specimens and curiosities when, in 1687, he was offered the opportunity to join the household of Christopher Monck, Duke of Albemarle, as the Duke's personal physician. Albemarle had recently been appointed Governor of Jamaica, and in taking advice from close friends as to whether he should accept the post, Sloane was clear of the opportunities it would afford him as a naturalist. As he explained to fellow physician and naturalist John Ray, 'next to the serving of his grace and family in my profession, my business is to see what I can meet withall that is extraordinary in nature in those places'. He added that the possibility of the Jamaica trip 'seem'd likewise to promise to be useful to me, as a Physician; many of the Antient and best Physicians have travell'd to the Places whence their Drugs were brought, to inform themselves concerning them'.[39]

Ray, marooned at home through ill-health, was in no doubt that plant taxonomy would benefit immeasurably if a naturalist of Sloane's calibre were prepared to make such a journey:

> Were it not for the danger and hazard of so long a voyage, I could heartily wish such a person as yourself might travel to Jamaica, and search out and examine thoroughly the natural varieties of that island. Much light might be given to the history of the American plants, by one so well prepared for such an undertaking, by a comprehensive knowledge of the European. Nay (which is more), that history, we might justly expect, would not only be illustrated but much improved and advanced.[40]

'We expect great things from you', Ray continued, 'no less than the resolving of all our doubts about the names we meet with of plants in that part of America. You may also please to observe whether there be any species of plants common to America and Europe.' Martin Lister, who was compiling his illustrated *History of Shells* at the time, urged Sloane to go and to 'collect & transmitte hither ye land snailes & such shells as shall be found in ye Fresh water rivers or ponds of Jamaica which will very much oblige his most humble servant'. 'Also,' he added, 'to observe, whether there are any naked snailes in Jamaica, I meane such as are naturallie without shells at Land as with us.'[41]

Sloane was convinced. On 12 September 1687 he sailed from Spithead on the frigate *Assistance*, part of a small fleet that also comprised Albemarle's yacht and two large merchant ships. They were almost immediately forced back into port by atrocious weather, finally putting to sea on 5 October. The party reached Jamaica on 19 December 1687. There Sloane dedicated himself to assembling a vast collection of local specimens, as he describes in the preface to the *Natural History* he later published:

> After I had gather'd and describ'd the Plants, I dried as fair
> Samples of them as I could, to bring over with me. When I
> met with Fruits that could not be dried or kept, I employ'd the
> Reverend Mr. Moore, one of the best Designers I could meet
> with there, to take the figures of them, as also of the Fishes,
> Birds, Insects, &c. in Crayons, and carried him with me into
> several places of the Country, that he might take them on the
> place [*in situ*]. When I return'd to England, I brought with me
> about 800 Plants, most whereof were New, with the Designs
> before-mentioned, &c. And shew'd them very freely to all lovers
> of such Curiosities.[42]

Sloane preserved large numbers of specimens, pressing the plants and smaller insects between large sheets of brown paper, preserving larger, fleshier items in alcohol. (His fellow collector James Petiver issued a sheet of instructions to would-be specimen-collectors, advising them to preserve 'small animales, as beasts, birds, fishes, serpents, lizards, and other fleshy bodies capable of corruption' in 'Rack, Rum, Brandy, or any other Spirits; but where these are not easily to be had, a strong Pickle, or Brine of Sea Water may serve'.)[43]

None of the large live specimens he attempted to keep to ship home survived. An alligator, which he kept in a tub of brine, died, while a 'large yellow snake' met a similar fate:

> Though I foresaw the Difficulties, yet I had an Intention to try to bring with me from *Jamaica* some uncommon Creatures alive, such as a large yellow Snake, seven Foot long. I had the Snake tam'd by an *Indian*, whom it would follow as a Dog would his Master, and after it was deliver'd to me, I kept it in a large earthen Jarr, such as are for keeping the best Water for the Commanders of Ships, during their Voyages, covering its Mouth with two Boards, and laying Weights upon them. I had it fed every Day by the Guts and Garbage of Fowl &c. put into the Jarr from the Kitchen. Thus it liv'd for some time, when being weary of its Confinement, it shov'd asunder the two Boards on the Mouth of the Jarr, and got up to the Top of a large House, wherein lay Footmen and other Domesticks of her Grace the Dutchess of *Albemarle*, who being afraid to lie down in such Company, shot my Snake dead.[44]

On 6 October 1688, a mere year after Sloane and the Duke of Albemarle had set out from England, Sloane's patient, who had not been in good health for some time, died. Sloane's last duty as Albemarle's personal physician was to embalm the Duke's body for shipment to England. Thus Sloane's skills as a preserver of 'fleshy' specimens were put to particularly good use (Lady Albemarle kept Sloane on for four years after their return, in recognition of the services he had rendered her husband). Embalming turned out to be a necessary expedient, since the departure of the dead Duke and his

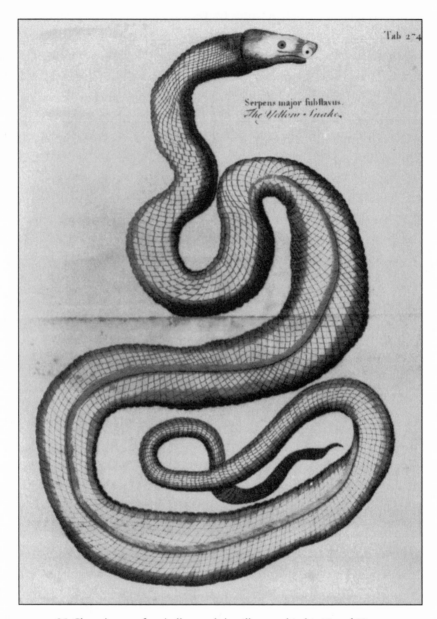

21. Sloane's seven foot 'yellow snake' as illustrated in his *Natural History*

household was held up for a further five months while the political situation in England became clearer (William of Orange landed in England with a military force in November, and was declared King in James II's place the following February). Eventually, on 16 March 1689, they set sail for home, and shortly after their return the Duke's body was buried in Westminster Abbey.[45]

Sloane never embarked on a comparable adventure again. But he continued for the rest of his life to enlarge his collections, while he pursued a highly successful career as a society physician. He had certainly profited from his participation in the successful, speculative wreck-raising venture in Jamaica shortly before Albemarle's death (Albemarle's widow lived royally on the proceeds for some years afterwards). Sloane enhanced this financial cushion for his collecting with the proceeds of several extremely profitable business ventures, which he embarked on as a direct result of the botanical work he had begun in Jamaica. The most successful of these was his adaptation to European use of a new therapeutic plant: cocoa, the source of chocolate.

Sloane describes in his *Natural History* how he stumbled on the idea of drinking chocolate, the 'milk chocolate' that was marketed over his name. In Jamaica, Sloane found, chocolate was regarded as having considerable therapeutic value for the digestion:

> Chocolate is here us'd by all People, at all times. The common use of this, by all People in several Countries in *America*, proves sufficiently its being a wholesome Food. The drinking of it actually warm, may make it the more Stomachic, for we know by Anatomical preparations, that the tone of the fibres are strengthened by dipping the Stomach in hot water, and that hot Liquors will dissolve what cold will leave unaffected.

However, drunk mixed with honey and pepper, as was customary in the West Indies, Sloane found the beverage quite unpalatable to a European:

I found it in great quantities nauseous, and hard of digestion, which I suppose came from its great oiliness, and therefore I was unwilling to allow weak Stomachs the use of it, though Children and Infants drink it here, as commonly as in *England* they feed on Milk. Chocolate colours the Excrements of those feeding on it of a dirty colour.[46]

'As commonly as in England they feed on Milk' turned out to be the clue. Sloane financed the commercial production of chocolate in blocks, and sold it with instructions that it should be mixed with hot milk and drunk in the form of the delicious beverage still consumed in quantities in Britain today, 'for its Lightness on the Stomach'. Sloane's imported remedy set a fashion (Hooke records in his diary when and how much chocolate he consumes, and how it affects his health), and rapidly became a consumer delicacy, joining

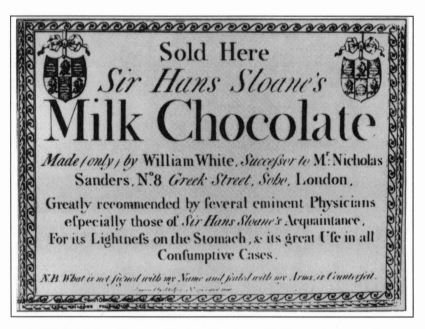

22. Label from Sloane's milk chocolate, from which he made a personal fortune

that other therapeutic drink, coffee, as the preferred beverage of those *à la mode* in London intellectual circles.

Chocolate, marketed as a remedy for stomach ailments and consumption, together with other expensive imported remedies, like Peruvian quinine bark, marketed as a remedy for 'tertian ague' (malaria, which was widespread in England during the summer) made Sloane a sizeable fortune. He joined the ranks of wealthy London merchants like James Petiver and Peter Collinson who were in a position to be thoroughly proactive about their passion for natural history and collecting – funding expeditions to North America and the West Indies in search of ever more unusual specimens, exploiting their privileged access to the trading routes plied by their own merchant vessels. They collaborated with one another in these expeditions, as well as in procuring desirable additions to their cabinets by purchasing from existing collections of high repute. They left their collections to each other in their wills, thus ensuring that continuing care was taken of them, even after their deaths.

These businessmen-collectors were well aware that their expensive 'hobby' doubled as an efficient means of locating luxury commodities that could be exploited commercially. Sloane became eager to acquire the German physician Engelbert Kaempfer's collection of Japanese curiosities in 1712 when he read Kaempfer's *Exotic Pleasures*, an account of life in Persia and Japan (he did indeed purchase Kaempfer's curiosities and books after his death). A section of this work describes (with helpful illustrations) a range of Oriental remedies and intoxicants, including the Japanese tea plant, 'dragon's blood' (a plant preparation), the white opium poppy and cannabis; Kaempfer explains how these are prepared and used therapeutically (tea is described as cleansing the blood, and washing away the 'tartarous matter of calculous concretions' of gout). These were the 'designer drugs' of the period, and demand for them grew steadily.[47]

A visitor to Sloane's collection in 1710 reported that he had seen 'the whole Courten collection' there. Courten (who sometimes called himself Charleton) had left his extraordinary collection (valued at some £50,000) to Sloane in 1702. Before that, it was Sloane who had brought back specimens specially for Courten from his Jamaica voyage: 'I gave my very particular and intimate Friend Mr. Courten whatever I brought with me, that he wanted in his extraordinary Museum,' he wrote in his *Natural History*. The same visitor recorded that Sloane was 'daily increasing his objects in England for vast sums of money'. In 1711 Sloane asked James Petiver to go to the Dutch Republic on his behalf to purchase Paul Hermann's renowned collection of Cape plant specimens and illustrations (Hermann had died in 1695, and his collection was now on the open market). Before Petiver set out on this, his only foreign excursion, he wrote to Sloane that 'in case I should dy before my return from Holland, I make you sole possessor of all my collections of naturall things whatsoever'. No disasters befell him, however, and he informed Sloane that he had 'bought [him] the greatest share of the choisest' of Hermann's specimens – a surviving lading document for the ship on which he returned reveals that he brought back 'three cases and one casket' of botanical materials.[48] At Petiver's own eventual death in 1718, he left his collection to his sister, from whom Sloane bought it intact, for the sum of £4,000.[49]

By 1722 the diplomat and botanist William Sherard was complaining somewhat plaintively to a friend that Sloane, 'wallowing in money', had probably already sewn up the acquisition of specimens coming from Carolina (the outcome of an expedition that the two men had both helped fund): 'I cannot think he has been unsuccessful, having had a large share of all that has come into England, and I never yet had a single Plant.'[50]

At the end of his life, Sloane had siphoned up practically every major botanical collection, 'dried garden' and cabinet of curiosities whose owner was prepared to relinquish it, either for cash or as a bequest to what had become the dominant collection in the land. So considerable were his holdings in some areas that they verged on completeness, according to the judgement of contemporary experts, thus becoming invaluable systematic collections in their own right. Ray told Sloane, for instance, that his collection of insects outclassed all the others he had inspected – 'there are not many species wanting that are in other men's hands' – but, then, Ray was by this time dependent on Sloane for medical advice and financial support.[51]

Thus was the extraordinary collection of rarities and curiosities of every kind assembled that at Sloane's own death became the basis for the British Museum Collection. Under the terms of Sloane's own will, his entire collection was left to the nation (or, more precisely, to London), for 'the use and improvement of physic, and other arts and sciences, and benefit of mankind'. He had hoped that it might be housed near the Chelsea Physic Garden (whose control he had acquired under one of his many land deals, in 1722), but in the end a home was found for it 'whole and intire' at Montague House in Bloomsbury – formerly the home of Ralph Montague, and designed and built, appropriately, by Robert Hooke (though substantially rebuilt to his design after a fire destroyed the original building).[52]

The founding collections of specimens in the British Museum are thus the product of one man's idiosyncratic interests, expanded more or less haphazardly as he acquired the contemporary collections of others that were deemed 'important' (and to which Sloane's ample means gave him almost unrestricted access). Which certainly raises the question as to what kind of guiding principles shape the museum and its contents as they have come down to us. A repeated refrain, around 1700, is that collections of rarities (and, indeed, collections of papers, and libraries) are left by their owners uncatalogued and in no systematic arrangement. What this points towards is a fundamental difficulty intrinsic to taxonomies in terms of completeness. Every arrangement of data is structured according to the presiding organiser's personal agenda, whether this is made explicit or not. As soon as that individual is no longer available to impose their 'shape' on the project, there is a sense that the whole thing is escaping from order.

After Robert Boyle's death, his vast accumulation of personal papers and manuscripts were described by the three colleagues who undertook to catalogue them as being in a state of total disorder. They found it impossible to agree on what principles to adopt in order to 'organise' them.[53] Boyle's own flexible method of archival storage seemed chaotic once the presiding, organising mind had gone. Worse still, those like Boyle's biographer Birch, who did feel able to impose some kind of meaningful shape on the Boyle corpus, had a set of guidelines of their own for what constituted 'scientific' study of the first order. Accordingly, they purged the collection of anything 'extraneous', which failed to fit, thereby producing a convincing memorial to the 'father of modern chemistry', at the expense of lifelong interests of Boyle's such as alchemy, evidence of which was either destroyed or left out of the picture.[54] In the case of the British Museum too, the trace of the idiosyncratic Sir Hans Sloane has all but disappeared, the personal collector separated from his collections for ever.

·CHAPTER 7·

Examining the Evidence

Meanwhile, back home

THE ROYAL SOCIETY in London received a steady stream of unusual specimens of minerals, rocks and other curiosities from overseas travellers, merchants and sea-captains, as gifts. They were presented, either by the Curator of the day or by a friend or patron of the donor, at the Society's weekly meetings for discussion. The minutes of meetings at the Royal Society for this period record oddly inconsequential discussions arising from these sessions, wandering from the precision of pure mathematics to the quirkiness of personal remedies for minor ailments, with 'show and tell', and all kinds of hearsay and fantastic hypotheses in between. These were the result of the encounter between curious minds and rapidly expanding horizons of knowledge, as overseas colonisation stimulated long-distance acquisitiveness, and the global gathering of scientific materials.

In December 1682 a letter from the Dutch physician Willem Ten Rhijne, concerning therapeutic plants in Java, was read at a Royal Society weekly meeting. It provoked a wide-ranging discussion, in which one member reminded his colleagues that the

camphor tree growing in the Chelsea Physic Garden had been sent
to England by Ten Rhijne, and that Dutch specialist gardeners had
succeeded in growing tropical plants, including cinnamon and
nutmeg at Groningen in the north of the Dutch Republic.
Discussion then passed to Oriental medicinal practices in general,
and their efficacy in treating widespread ailments like gout. As a
result of the interest shown, the Royal Society took a decision to
undertake publication of Ten Rhijne's work on Asian medical practices the following year (printed for the Society on presses located
in the basement of Wren's Sheldonian Theatre in Oxford). Ten
Rhijne's book includes the first European account of acupuncture
(including cupping) and moxibustion (herbal poultices called moxa
burned on to the skin).[1] The author had served as a Dutch East
India Company physician in Japan, where acupuncture and burning
moxa were widely practised.[2] Following Ten Rhijne, Thomas
Sydenham, another Royal Society member and a rising society
physician in London, took up the subject of moxibustion in his
own *Treatise on Gout* (1683).[3]

In 1689 Sloane brought back quantities of unusual materials
from the West Indies and North America, in packing cases stowed
on board ship alongside the embalmed body of his ex-employer,
Albemarle. Most of this material he kept and classified for his own
collection. Some of it, as he acknowledged in his *Natural History*, he
supplied to established collectors like Courten, whose collection he
greatly admired, though later, when he acquired it himself, he complained at its lack of organisation.

He made yet other materials available to individual Royal Society
scientists, as colleagues and friends, for particular experimental
programmes in which they were engaged. Robert Boyle, thirty years
older than Sloane, was one of the scientific practitioners Sloane
admired and strove to emulate. He cultivated their friendship 'by
communicating to him whatever occur'd to himself, which seem'd
curious and important, & which Mr Boyle always receiv'd with his

usual Candour & return'd with every Mark of Civility and Esteem'.[4] Accordingly, on his return from Jamaica, Sloane provided Boyle with unusual materials for experiments acquired in the West Indies.

At the time, Boyle was conducting comparative experiments on the specific gravities of natural substances as a way of checking the richness, or density, of metal ores (essential to merchants purchasing on site in places like West Africa). He was also testing the hypothesis that specific gravity of naturally produced materials varied according to geographical location. In 1690 Boyle tabulated the relative specific gravities of unusual, naturally occurring stones, including bezoar stones (solid lumps of vegetable matter accumulated in an animal's intestinal tract, either evacuated or retrieved at autopsy).[5] He found that a 'fine Oriental' bezoar stone had a higher specific gravity than one obtained locally; a specimen of 'lapis manati' had a significantly higher specific gravity than 'another from Jamaica'. A specimen of Venetian talc (the stone from which dusting powder is obtained) had a lower specific gravity than a Jamaican

I. Drawer of mineral specimens including bezoars from Sloane's collection

one. The Jamaican specimens were either acquired permanently by Boyle from Sloane or, more likely, they were loaned for the purpose.[6] (Sloane later stopped giving fellow scientists free access to his specimens, when he found that some of his colleagues were 'so very curious, as to desire to carry part of them home with them privately, and injure what they left'.)[7]

Sometimes the exchange of materials and ideas went in the reverse direction, from practising scientist to collector. Sloane was interested in fossils and geological specimens, and was among those, like Hooke, who believed on the basis of their inspection that fossils were prehistoric remains of living things. When Edmond Halley had returned from his astronomical expedition to St Helena in 1676 he had brought back with him geological specimens that had caught his eye. Sloane added one of these to his extensive collection of rock samples, and attached a note on its provenance, and a comment on its possible geological origins: 'A piece of red stone or mineral from St Helena seeming ironstone in which are lodged some silver mica or talc. It looks as if it had been a volcano cinder. Dr Halley saith the most part of the island is of it & believes it cast up from the sea bottom.'[8]

In France, generous financing of expeditions by Louis XIV (via his finance minister Colbert and his successors) provided the Paris Académie with a comparable fund of rare specimens for experimental purposes. Jean Richer brought botanical materials back from his Académie-funded voyage to Cayenne. Before they set out for China, Tachard and his Jesuit colleagues visited the Académie 'to learn what matters of natural history the Academy would like them to correspond about', and what kinds of specimens they should bring back with them. The Académie received a paper on the medicinal value of ginseng from a diplomatic traveller to the East. Just like their English counterpart, the Académie absorbed a steady influx of overseas material and information into its own research and publications.[9]

The impact of such imported knowledge and materials on scientific thinking was further amplified by the regular exchanges between the two institutions, which continued largely unabated even during periods of high political tension between the English and French governments. Among those encountered in the present story, Christiaan Huygens visited the London Royal Society in 1661, 1663 and 1689, where both Huygens and Gian Domenico Cassini were made Fellows. Edmond Halley, Martin Lister and John Locke visited the Paris Académie; all of them also exchanged scientific information by letter with their counterparts across the Channel. Beyond actual visits, Henry Oldenburg brokered extensive correspondence between the members of the two academies, and supplied English scientific publications to the French, hot off the presses, and vice versa.[10] Such exchanges produced common topics of discussion, based on shared data and freely referring to specimens and reported phenomena to which both scientific communities had access.

Cash transactions had their part to play alongside this variety of intellectual exchanges. Unlike Halley, Boyle and Sloane, Maria Sybilla Merian did not indulge in genteel exchange of interesting samples and experimental materials with other like-minded people. When she returned from her two-year stay in the South American colony of Surinam in September 1701, she was in poor health and serious financial difficulties. Fortunately, there was a growing number of collectors prepared to pay for rarities to enlarge their collections. Merian supported herself and her daughters by advertising both watercolours and preserved specimens she had brought back with her for sale on the open market.

Merian was reluctant to part with her original paintings, on the grounds that although they would 'be worth the money and the expenses of the Surinam journey due to their great rarity, then only one person would have them' (whereas publication would make them available to many). However, in October 1702 she offered a

Nuremberg collector a wide range of Surinam insects, reptiles and small birds:

> I have also brought with me all the animals comprised in this work, dried and well preserved in boxes, so that they can be seen by all. I also currently still have in jars containing liquid one crocodile, many kinds of snakes and other animals, as well as about twenty round boxes of various butterflies, beetles, humming-birds, lantern flies (referred to in the Indies as lute-players because of the sound they make), and other animals which are for sale. If the gentleman desires to have them, he need only so order.[11]

From early in 1703 Merian was in correspondence with the affluent businessman and collector James Petiver in London, who she hoped would subsidise publication of her Surinam volume (he was indeed helpful, and found other English subscribers). Later her younger daughter Joanna and her husband returned to Surinam, and together the three operated a profitable transatlantic business in unusual imported specimens. In 1712 Merian wrote to Petiver:

> I have received Surinamese insects from my daughter: a hey [shark?] and 4 small 'hayen', a large iguana and a small one, a Sauvegarde [lizard], a flying fish, a spider, 6 snakes, two thousand bones, 2 lizards, an animal that attaches itself to ships, a small fish, all in bottles filled with spirits, also two boxes containing 30 insects, which she wishes to sell for 20 Dutch guilders. If you wish to have everything delivered, please place your order.[12]

We do not know whether Petiver responded to this invitation to purchase any or all of these items; however, Petiver's large and diverse natural-history collection, which Sloane bought from his sister after Petiver's death, included original Merian watercolour drawings.

In these ways, the whole landscape of intellectual markers and

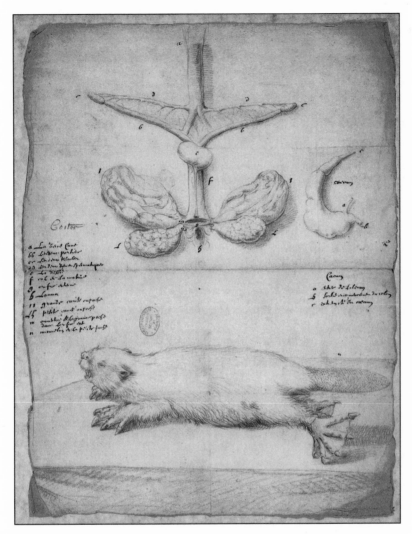

2. The French academician Perrault's red chalk drawing of a beaver and
its organs from his *Natural History of Animals*

reference points within which the scientific community operated,
broadened to include the furthest reaches of East and West. Men and
women who had never left London could soon recognise a South
American mineral fragment, an Indonesian shell, or a blossom from
the African Cape.

To be taken in small doses

Innovative medical treatment from Asia entered the world of the London virtuosi and their fashionable remedies almost by accident, in the descriptions that accompanied the gorgeous images of strange stones and botanical rarities. 'Peruvian bark', whose use in the treatment of malaria by Amerindians was described by European travellers to South America, caused a particular stir. It quickly became the remedy of preference among discerning doctors for anything resembling malaria's recurrent fevers. Dr Richard Morton, in his published treatise on medical diagnosis, reported that the poet Andrew Marvell (who had died in 1678) had been killed by an overdose of opium, administered by a doctor who was not prepared to try this new miracle remedy as the treatment for the acute intermittent fever from which Marvell was suffering.[13] By the end of the century, Morton's view of *cinchona* (as it was also known) as a wonder drug for the treatment of a whole range of malaria-related, intermittent fevers endemic in Europe was shared by most scientifically minded doctors. Thomas Sydenham was of the opinion that Peruvian bark alleviated the symptoms of malaria, rather than actually offering a cure, but he regarded it as the only remedy worth administering nonetheless. Thomas Willis agreed – which is not entirely surprising, since Sydenham, Willis and Locke conducted rudimentary field-trials together on *cinchona*, recording its effect on their patients on specially produced questionnaires.[14]

Cinchona, from which modern quinine is derived, was probably the most impressively successful of the new medical materials discovered by botanists travelling to remote parts of the globe in the seventeenth century. Jesuit missionaries in Peru had observed native use of the bark for 'agues' or fevers, and brought it back to Europe – hence *cinchona* was also known as the 'Jesuit bark'. Hence also the fact that those who disapproved of this 'new-fangled' drug described it as a popish imposture, and its medicinal effects as the work of the

Artemisia Jul. 23   Common Mugwort   515

3. Specimen of mugwort from Locke's personal herbarium. The red
stalks supposedly indicated its efficacy for curing peculiarly female
disorders

devil (there was an apocryphal story that Oliver Cromwell had refused it on his deathbed on precisely these grounds). 'Since *per se* it neither operates by Vomit, Stool, Urine, or Sweat,' wrote one physician opponent, 'we may safely conclude its chief Energy consists only in stopping the Ague fits, whereby worse Diseases are engendered.'[15]

Respectable botanists, however, were prepared to vouch for *cinchona*'s genuinely curative properties. John Ray devoted a chapter of his *History of Plants* to the Peruvian tree from which the bark came, and reported that Sydenham had identified it as a great 'specific' (that is, a specialist drug which specifically cured malarial fevers, not an all-purpose remedy), on the basis of large doses administered to patients, instead of the excessively small doses used when the powdered bark first arrived in the country.

Ray's confidence in the Peruvian bark may have had something to do with his high regard for and close friendship with Sir Hans Sloane. In his *Natural History*, Sloane recorded using Peruvian bark in the treatment of a range of illnesses while he was in Jamaica. On his return he 'invested the greatest part of the Fortune he acquired there' to set up a business importing the bark, and was naturally inclined to advertise its virtues as widely as possible.[16] He advised balancing the purging quality of the bark by mixing it with 'a very little quantity of Laudanum [opium water]'; children should take it 'in Chocolate well sweetned'.[17]

It was probably as a result of Sloane's encouragement that his colleagues began to extol the medical use of *cinchona* for the treatment of frequently occurring conditions other than the intermittent fevers for which it had proved genuinely therapeutic. Locke thought it might be helpful for the treatment of gout:

> Gout is a disease of malnutrition, which gradually but slowly
> accumulates and finally issues in paroxysms like those of the
> intermittent fevers; the cure therefore seems to be by Peruvian

Bark and other medicines which help the digestion and
assimilation of the food.[18]

Sydenham's prescription for gout was strikingly similar, with the
addition of opium to provide relief during acute attacks: the patient
should have 'immediate recourse to laudanum, twenty drops of it in
a small draught of plague-water', then, 'of all simples the *Peruvian
bark* is the best, for a few grains taken morning and evening
strengthen and enliven the blood'.[19]

These physicians were all members of the Royal Society, frequenting the same circles as Hooke, Halley and Boyle. They
shared a preference for medical materials that had been seen in use
in their native habitats, fully identified and described botanically,
then tested in rudimentary clinical trials at home, over any remedies suggested by high-blown theoretical medicine. Sloane wrote
up the sensational curative powers of *cinchona* for the official journal of the Royal Society, with an enthusiasm that perhaps smacks
of marketing hype:

> It overthrows with one simple Medicine, without any
> preparation, all the Hypotheses, and Theories of Agues, which
> were supported by some Scores not to say Hundreds of
> Volumes, which 'tis plain did mischief by hindering the
> advantage Men might have received sooner from so innocent
> and beneficial a remedy.[20]

Of course, Sloane himself was the adventurous physician *par excellence*, who had actually been prepared, in his youth, to undertake
pharmaceutical fieldwork in Jamaica, and he remained committed to
exotic remedies. In his own collection of medical materials, Sloane
included a specimen of 'bamboo tarr', which he labelled as follows:

> Bamboo *tarr* an ointment used for the cure of the gout & the
> rheumatism in Guinea [West Africa]. This I believe is a
> substance from the Duke of Chandois calld *Unguentum Encoo*

or the white balsam, which is the excrescence of a tree
growing in the country of Entan lying six days journey inland
from Ashanta being to the eastward, it is also procured at
Agrafar & Creepu Country to the Eastward of the river
Volter & likwise at Whidah. The blacks made use of this
ointmt. to anoint themselves in place of tallow being much
preferable thereto & extremely good for all manner of aches,
paines & sores.[21]

No doubt, had bamboo tar achieved a vogue among malaria and
gout sufferers in England comparable to that of the Peruvian bark,
Sloane would have 'procured' and marketed it in the same way.

By contrast with Peruvian bark, rhubarb root was not unknown
to European medicine prior to the great expansion in the accessible
world in the seventeenth century. Gentle laxative and emetic remedies, which purged without irritating the bowel and intestines, had
always been highly regarded by both Eastern and Western physicians – remedies that supposedly emptied the body of the noxious
substances causing its sickness. Dried rhubarb root had figured
since antiquity among medical material obtained from the Far East,
and was mentioned by both the ancient Greek medical authority
Dioscorides and the Roman naturalist Pliny, as a particularly mild,
yet effective purgative.

The finest medicinal rhubarb came from China. It had reached
Mediterranean Europe by the twelfth century, introduced by Arab
pharmacists along with other mild laxatives and cathartics, like
cassia and senna. In 1295, Marco Polo found an abundance of
rhubarb growing in south China.[22] By the seventeenth century
large amounts of dried rhubarb root were arriving in Western
Europe, either directly, or via Russian merchant intermediaries.
From the 1650s onwards, the Russian trade in rhubarb imported
from China, via Siberia, was state-controlled, and produced an
extremely profitable export to Western Europe. Meanwhile, Russia
built up a thriving rhubarb-root production of its own in southern

Siberia, cultivating a variety of the plant that was native there (though this was acknowledged to be a somewhat inferior variety, medicinally).[23]

Purging was the corner-stone for all seventeenth-century medical practice. In his *Rational Pharmaceuticals* (1674) Thomas Willis gave his explanation of how purges worked. Cathartic medicines adhered to the nervous fibres of the stomach, saturating and irritating them until 'the viscous Phlegm laid up in the folds of the Stomach' forced its contents downwards towards the pylorus, to be eventually expelled. Particles of the medicine also entered the blood, similarly forcing unwanted materials towards the lower intestines, cleansing and purifying the bloodstream. What made 'true rhubarb' so desirable was the fact that at its best – in the form of roots grown and properly prepared in Asia – it produced frequent, but not unpleasant, expulsion of stools, without stimulating an accompanying vomiting. Having once produced evacuation, it apparently 'left a very astringent binding quality behind it', and could therefore be used even in cases of fluxes, dysenteries and diarrhoeas.[24]

The goal for seventeenth-century European botanists was to cultivate Chinese rhubarb successfully at home. Experience had already shown that able horticulturalists in Amsterdam, or the Chelsea Physic Garden, could establish plants native to Asia, India and the Americas in their botanical gardens. Anyone who could produce such an expensive medical material as 'true rhubarb' locally was clearly in line to set up a profitable business in home-grown pharmaceuticals. But 'the very true rhubarb' proved curiously elusive – through the whole course of the seventeenth century able botanists failed even to identify correctly the 'true China rhubarb', let alone grow it. Instead, a series of lesser rhubarbs and rhubarb-like impostors were assiduously cultivated from seed, producing roots and rhizomes that singularly failed to meet the medicinal standards of the expensive imported variety.

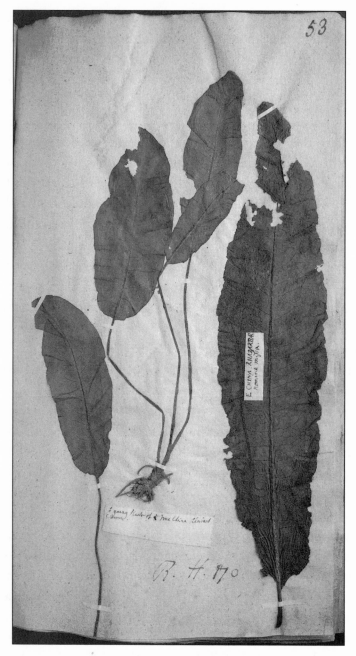

4. Specimen identified in error as true rhubarb from Sloane's herbarium

It was the influential and widely connected James Petiver who provided Sir Hans Sloane with two plants of 'the true China rhubarb' for his herbarium, as part of a consignment of unusual Chinese medical materials he had acquired from an English surgeon with the East India Company in Madras (who had, however, never visited China himself). Sloane was delighted with specimens from such a reliable source, recording that 'both were brought me lately from China after I had before many times been disappointed'. He duly pressed and mounted them, labelling one, 'a young plant of the true China Rhubarb', and the other 'East China *Rhabarbar*'. Neither, however, is actually rhubarb at all (one is clearly a plant of common dock). Given Sloane's and Petiver's considerable botanical expertise, the misidentification of these specimens is a clear indication that no one in London had ever actually seen living (let alone growing) examples of the plant that yielded the dried root, or dramatically yellow powder, of China rhubarb.[25]

The confusion over 'true rhubarb' was not resolved until the nineteenth century. Meanwhile, root material from plants grown in Europe's botanical gardens and private yards continued to disappoint the medical community, and the East India Company and the Russian State continued to make immense profits by importing and distributing a highly desirable commodity that was in constantly short supply. In addition to direct prescription by physicians, rhubarb was a key ingredient in any number of 'over-the-counter' lozenges. A typical rhubarb pill of the period contained ten drams of the best rhubarb root, finely ground, mixed with the juices of agrimony and bitter almonds (four drams each), four drams of rose leaves, and small amounts of spikenard, anise, madder, wormwood (absinthe) asarabacca and smallage-seed. These pills were supposed to cure all 'depraved' actions of the liver, and to prevent incipient dropsy and jaundice.[26]

Morbid interests

Remedies like these catered for an eager clientele of affluent men and women, for whom the unrelieved physical discomfort of minor ailments was a fact of life, and who knew that an infected cut or a boil might at any moment develop into a grave medical condition. On 30 October 1672, Dr John Wilkins, Bishop of Chester, presided at a weekly meeting of the Royal Society. The meeting was an eventful one, packed with highly varied items of scientific interest: Hooke, Wilkins's protégé and devoted friend, recorded in his diary that Ashmole's *History of the Order of the Garter* was presented to the members, also Hevelius's observations of the recent comet, as well as scientific materials compiled by the Italian biologist Malpighi, the observations of the recently deceased young astronomer Horrocks, and a map of Nova Zembla (an Arctic island off the north Russian coast), introduced by Oldenburg.[27]

Just two weeks later, to the great distress of his circle of friends and fellow scientists, Wilkins lay dangerously ill. 'Lord Chester desperately ill of the stone, stoppage of urine 6 dayes,' reported Hooke. On 16 November Hooke made a note of the medical remedies that a sequence of distinguished medical visitors had prescribed Wilkins: 'oyster shells 4 red hot quenched in cyder a quart and drank, advisd by Glanvill. Another prescribed *flegma acidum succini rectificatum cum sale tartari* [acid precipitate of amber [?] refined with salt of tartar]. Dr. Goddard advisd Blisters of cantharides [poultices of ground-up beetles] applyd to the neck and feet or to the vains' – all to no avail. On 18 November Hooke was 'at Lord Chesters, he was desperately ill and his suppression continued'. Next day he wrote in his diary that, 'Lord Bishop of Chester dyed about 9 in the morning of a suppression of urine.'[28]

However, this was not quite the end of the story. The following day an autopsy was carried out on Wilkins's body, the results of

JOAN.WILKINS PRIMUS SOCIETATIS REG:SECRET:ET BEN EFACTOR WADHAMENSIS GARDIANUS DEC:RIPON:MAGISTER COLTRIN:CANT:EPISCOPUS CESTRIENSIS OBII

5. John Wilkins, Bishop of Chester, in good health.

which one of the doctors concerned quickly reported back to Hooke, who no doubt circulated the information among those who shared the fashionable coffee-houses he and his intellectual friends frequented, which provided the gossip-circulating service of our tabloid newspapers. Apparently there had been nothing physically wrong with Wilkins, in spite of his painful and protracted inability to urinate: 'Dr Needham brought an account of Lord Chesters having no stoppage in his uriters nor defect in his kidneys. There was only found 2 small stones in one kidney and some little gravell in one uriter but neither one big enough to stop the water.' The word on the street was that Wilkins had died of the side effects of the medicines he had been so enthusiastically prescribed: "Twas believd his opiates and some other medicines killed him, there being noe visible cause of his death,' reported Hooke.[29]

In Hooke's view, the treatments prescribed by contemporary physicians were as likely to kill as to cure (he records similar comments regarding the death of Sir Robert Moray in 1673, whose autopsy also revealed his body 'very intire and sound and nothing amisse, and noe visible cause of his death').[30] On the other hand, Hooke was entirely committed himself to a therapeutic regime of drug-taking. He believed this would not only alleviate the symptoms of a whole host of minor complaints, but would also prevent the onset of gout and the stone, which carried off so many of his friends and acquaintances.

Indeed, the diary Hooke kept for the period 1672 to 1680 is first and foremost a record of his day-by-day regimen and drug therapy. He records precisely what has entered his body — foods he has chosen himself for their health-improving properties, refreshments he has been served at the tables of others or has taken at the coffee-house, and the purgatives and other medicaments and remedies he has taken, morning and night. He then describes his bodily symptoms, and any 'cleansing' consequences (excreting, vomiting or urinating) that have ensued. He also specifies how he feels 'in himself' at the end of the process. He embarks on a 'course' of, say, *sal ammoniac*, and comments on the outcome. He observes the regimens of others, especially when their diet proves fatal (as in the cases of Wilkins and Moray).

Some of the treatments Hooke tries are exercises in medical self-experiment. That is to say, he examines the consequences of ingesting particular, highly rated remedies, by trying them systematically on himself. In this he is probably fairly typical, though few private records as meticulous as his diary survive. In the course of their correspondence with one another on all kinds of experimental matters, scientific contemporaries of Hooke regularly describe to one another symptoms of their diseases, and treatments they have already tried. They request prescriptions, suggest remedies to one another, and pass on all kinds of bits of folk medicine, whether they are physicians or not.

In France, glimpses of the practice of self-administering medicine experimentally are to be found more prominently featured. Physicians arguing for and against the therapeutic value of the metallic compound antimony (stibnite, or antimony sulphide), had publicly admitted trying it out on themselves. Jacques Grévin, a member of the camp hostile to the widespread medical use of antimony, for instance, reported:

> I wanted to experiment upon myself, as being a thing as easy to take as a powdered grain of wheat. So I mixed a mere three grains with a little conserve of roses, as a result of which, in less than an hour, there followed such a strange vomiting that although I am of my nature an easy vomiter, each time it took hold of me, I felt as if I were going to die. It took me thus eight times, and as many times it worked on me at the other end, as a result of which it left me as if out of myself, and greatly enfeebled, a state which continued for a good week. All that it purged me of was a watery matter, which I have also observed in several others who have taken antimony. And there is no doubt that the purging it effects is the same in the healthy, the distempered and the sick, except that the effect is modified by the mixture of some kind of humour, which is perhaps intermingled.[31]

From this Grévin concluded that antimony was not merely a purgative but poisonous, and certainly not a cure. A physician of the opposing faction, Georges Handsch, on the other hand, reported

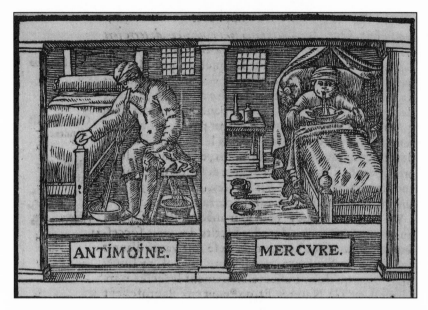

6. Purging with antimony and mercury

that he had himself taken antimony when he was suffering from nausea: he had taken three grains of antimony mixed with 'succre rosat' [conserve of roses], after which he vomited in less than half an hour, and his stomach upset had disappeared. Further evacuations improved his condition. He soon regained his appetite and gave thanks to God for his recovery from a dangerous illness by means of antimony.[32]

Hooke was prepared to swallow any number of potentially dangerous substances, producing violent and disagreeable reactions in the body, with alarming frequency. On I September 1672, for example, he 'drank steel' (a mixture prepared by quenching hot steel in wine or, alternatively, steeping a piece of steel in wine for several days). On 2 September, he 'tasted tincture of wormwood, eat raw milk, wrapt head warm and slept well after'. On the third, he took three ounces of infusion of *crocus metallicus* and vomited. (He also had an orgasm, which he regarded as part of his therapeutic regime, and 'slept pretty well'.) On the fourth, he purged seven times and reported feeling 'disordered somewhat by physic'. He also reported that his 'urine had a cloudy sediment, but brake not'. On the eighth and fifteenth respectively, he drank three pints of Dulwich water (a proprietary medicine laced with metallic compounds), which refreshed him. On the seventeenth, he took one pipe of tobacco. The entry for 22 September 1672 runs: 'Read Serlio's *Treatise on Architecture*, took syrup of popys [opium], slept little with sweat and wild frightfull dreams.' He took opium again on 29 September.[33]

Hooke went through cycles of attachment to, followed by disillusionment with, one remedy after another. In mid-summer, 1675, he began to take spirit of *sal ammoniac* (ammonium chloride) on a regular basis. A sequence of diary entries beginning at the end of July of that year runs:

> 30 July 1675: Took SSA [spirit of *sal ammoniac*] with small beer at Supper. Very feavorish all night but slept well. Strangely refresht in the morning by drinking SSA with small beer and sleeping after it, it purged me twice.
>
> 31 July 1675: In a new world with new medicine.
>
> 1 August 1675: Took volatile Spirit of Wormwood which made me very sick and disturbed me all the night and purged me in the morning. Drank small beer and spirit of Salamoniack. I purged 5 or 6 times very easily upon Sunday morning. This is certainly a great Discovery in Physick. I hope that this will dissolve that viscous slime that hath soe much tormented me in my stomack and gutts. *Deus Prosperat* [God will make it work].[34]

Shortly afterwards, however, Hooke stopped taking *sal ammoniac* altogether, either because of unpleasant side effects, which he failed to report, or simply because it had ceased to make him feel significantly better.

On the whole, Hooke tended to give each new remedy a trial period, during which he concentrated almost exclusively upon it, and noted carefully what else he ate and drank, in case his diet influenced the effectiveness of the chemical or metal compound. When he started writing his diary in August 1672, he had been drinking an iron and mercury preparation for a while. He then moved on to the steel drink. After that, for a time he depended almost entirely on Dulwich water, a mineral water taken from a natural spring at that location. He also used Andrew's cordial loyally, drank Mrs Tillotson's physick ale, took Dr Goddard's tincture of centaury, and Aldersgate cordial, ate conserves and flowers of sulphur, drank senna and jalap in various liquids, took *sal tartari* and *sal ammoniac*, and used a Mexican jalap called mechoacan. All of these remedies were purges of varying strengths. In addition, though not routinely, he took a variety of emetics. When these did not produce the desired result, he employed either a feather or a whalebone. To help him sleep he took opiates and put nutmeg into his nostrils.[35]

From our modern viewpoint we can see that this kind of regimen was potentially seriously damaging to health, and at the very least tended to produce physical symptoms of its own, which Hooke clearly does not associate with the medicine (which he assumes is curing such symptoms, not causing them). On 27 April 1674, for instance, Hooke 'took Dr. Thomsons vomit [a cordial containing antimony and tartrates]. It vomited twice. Purged 10 or 12 times.' Later that night he went to the coffee-house, then came home to find he could not sleep and that his arms were 'paralytick', he had 'a great noyse in my head', and 'some knawing at the bottom of my belly'.[36] He regularly reports 'a strange mist before my eye', which we now recognise as a symptom of mild poisoning with metal compounds.

Taking 'physick' could also have serious consequences for a person's lifestyle. A generation or so earlier, Sir Francis Bacon's brother Anthony had been warned by his noble patron that men who took too much physic were not considered reliable employment prospects (both Bacon brothers were lifelong takers of opiates, and Francis Bacon took a rhubarb purge once a week).[37] One reason for this is obvious: having taken a purgative, one had to stay close to home, in case it took effect unexpectedly. Pepys, who also followed a regular regimen of purges and 'clysters' (enemas), records that 'This day I stirred not out, but took physique and it did work very well.'[38] In his diary, Hooke records paying a visit to Sir Christopher Wren at home in 1673, 'he very sick with physick taken the day before'.[39]

There were, of course, benefits to be had from all this physic-taking. Hooke himself believed one of the virtues of his many mildly poisonous remedies to be their contribution to clear sight and incisive thinking. Having taken *crocus metallicus*, in May 1674, he pronounced himself 'refreshed by it' and, following an unsettled night's sleep, found his 'fancy very cleer'. Taking advantage of this, Hooke put in some work on water-clocks, quadrants and scoto-scopes.[40] In the preface to *Micrographia* he implies that, just as there are scientific instruments that enhance sense-perception (like micro-scopes), so there are indeed medicines that sharpen it:

> The Members of the Assembly [the Royal Society] have begun
> anew to correct all Hypotheses by sense, as Seamen do their
> dead Reckonings by Coelestial Observations; and to this
> purpose it has been their principal endeavour to enlarge and
> strengthen the senses by Medicine, and by such outward
> Instruments as are proper for their particular works.[41]

Laudanum-takers seem to have believed that their 'syrup of pop-pies' contributed to clear thinking. Thus on the occasion in September 1672 when the laudanum Hooke had taken made him sweat and gave him nightmares, he read an Italian treatise on classi-cal architecture and discussed it the following day with Wren. Hooke's friend and colleague Edmond Halley was so taken with the

clearheadedness and calming effect of an over-large dose of opium
he had ingested accidentally (because it had been insufficiently
mixed into a compound he had taken) that he read a paper on it to
the Royal Society in January 1690, describing how 'instead of sleep,
which he did design to procure by it, he lay waking all night, not as
if disquiet with any thoughts but in a state of indolence, and perfectly at ease, in whatsoever posture he lay'.[42]

Hooke's health was always poor. He tried new remedies with the
same vigour and enthusiasm with which he embraced new technology: on 6 February 1674, for instance, he records in his diary, 'At
Spanish coffee house tryd new mettall with Antimony, iron and lead.
At Shortgraves [another meeting place]. Tryd reflex microscope.'[43]

The scientist himself under the microscope

Hooke knew in meticulous detail what had gone into his body in
the way of medical preparations and nutrients. He focused his
microscope on what came out of it. And although microscopy did
not apparently provide the same kind of breakthroughs into the
structure and regularities of the body that telescopes yielded for the
heavens during the same period, self-examination nevertheless carried the physician's understanding extraordinarily close to discoveries
not fully understood for another two centuries.

One of the phenomena whose greatly magnified image Hooke
inspected and reproduced in *Micrographia*, was his own clouded
urine – urine that contained, in suspension, a chalky substance, or
'gravel'. Hooke announced that close examination revealed this
'gravel' to be crystalline:

I have often observ'd the Sand or Gravel of Urine, which seems
to be a tartareous substance, generated out of a Saline and a
terrestrial substance crystalliz'd together, in the form of Tartar,

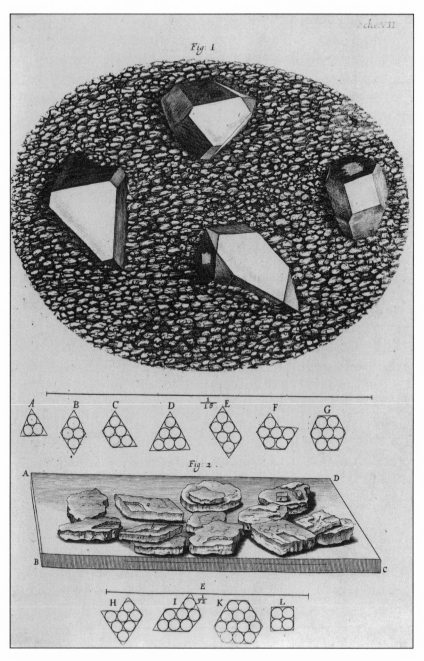

7. Hooke's observation under the microscope of 'gravel' in urine (fig. 2)

sometimes sticking to the sudes of the Urinal, but for the most part sinking to the bottom, and there lying in the form of coorse common Sand; these, through the Microscope, appear to be a company of small bodies, partly transparent, and partly opacous, some White, some Yellow, some Red, others of more brown and duskie colours.

Hooke found that the crystals could be made to dissolve again in 'Oyl of Vitriol [concentrated sulphuric acid], Spirit of Urine [impure aqueous ammonia, containing ammonium carbonate], and several other Saline menstruums'. Might not this offer a remedy for that most widespread of seventeenth-century complaints, 'the stone'? he suggested. What was required was a solvent for the crystals harboured by the blood, which could be injected directly into the bladder:

> How great an advantage it would be to such as are troubled
> with the Stone, to find some menstruum that might dissolve
> them without hurting the Bladder, is easily imagin'd, since some
> injections made of such bodies might likewise dissolve the
> stone, which seems much of the same nature.[44]

Hooke went on to suggest that Sir Christopher Wren's method for injecting liquids into the veins of animals, which Hooke hinted had been more widely tried than had been reported, by Boyle among others, might be adapted to introduce solvents into the bloodstream of those suffering from a variety of common ailments:

> Certainly, if this Principle were well consider'd, there might,
> besides the further improving of Bathing and Syringing into
> the veins, be thought on several ways, whereby several obstinate
> distempers of a humane body, such as the Gout, Dropsie,
> Stone, &c. might be master'd and expell'd.[45]

The crystalline structure Hooke had noticed in the deposits in his urine was also detected by his fellow-microscopist Antoni van

Leeuwenhoek in the chalky substances that seeped from the body of an acute gout-sufferer. In this case the sample examined was provided by an obliging relative suffering from acute gout, such that 'chalk came out of his joints', through a large hole in his heel, and a somewhat smaller one in his elbow.

'I asked him to let me have some of the chalk,' wrote Leeuwenhoek in one of his regular letters to the Royal Society, 'which he willingly granted me.' When he examined this 'chalk', Leeuwenhoek found that it was no such thing: 'I observed the solid matter which to our eyes resembles chalk, and saw to my great astonishment that I was mistaken in my opinion, for it consisted of nothing but long, transparent little particles, many pointed at both ends and about 4 "axes" of the globules in length.' Gouty deposits, like Hooke's 'gravel in urine', were made up of myriad tiny crystals. It was almost another two hundred years before the medical profession came to understand the significance of the structure of the deposits in gout – an excess of uric acid in the blood producing deposits of sodium urate in the joints – and moved towards a genuine cure.[46]

There is a certain irony in Hooke and Leeuwenhoek – neither of whom was a trained physician – coming so close, by simple experimental observation, to understanding that in all likelihood a high intake of lead, antimony and other metallic substances caused (rather than prevented) sickness. Within the arsenal of new medical plant materials and metal-based remedies taken as 'physic', some at least probably helped to produce the conditions the drug-taker sought to alleviate therapeutically, notably kidney stones and gout. One obvious culprit was the immensely popular *vinum emeticum* – emetic wine, or wine 'fortified' with tartar or nitre compounds. The Paris Medical Faculty itself endorsed these fortified wines as 'good for the health'. Even today, the label of any market-brand French mineral water carries a list proudly proclaiming the traces of a whole range of metals, supposedly good for the digestion.[47]

There is a further irony in the fact that in Hooke's own day the

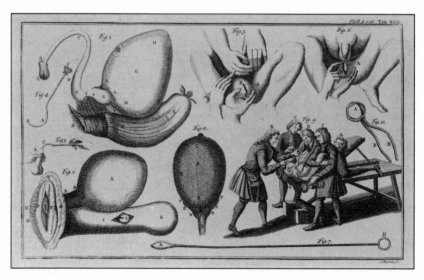

8. 'Cutting for the stone'

powerful College of Physicians was apparently more interested in protecting the entitlement of its members to prescribe the newly discovered therapeutic substances than in understanding the origins of the conditions from which their patients suffered. The Dutch physician Joannes Groenevelt pursued a successful career in London in the 1670s, operating ('cutting') to remove bladder stones. There were no objections until he began prescribing pills containing ground cantharide beetles as a diuretic, to purge stones without resorting to extreme surgical remedies. A former woman patient accused him of administering known poisons, the College of Physicians intervened on her behalf, on the grounds that Groenevelt was not licensed to prescribe, and Groenevelt was ruined.[48]

Putting things in order

From late middle age, the botanist John Ray, too, endured constant discomfort from unpleasant illnesses. He at first successfully dosed the 'herpes and sores' that began to trouble him at the age of sixty (in 1687) with 'hollyhock leaves boiled in butter'.[49] By the 1690s, when the sores on his legs became more painful and refused to heal, he told Sloane that his physician had given him 'a decoction of litharge of his own preparing to bathe them; I have taken flowers of brimstone [sulphur] inwardly, and applied an unguent to the soles of my feet.' By return of post, Sloane made his own suggestions. Ray reported that he was now taking 'inwardly', 'flowers of sulphur, half a drachm at once, which keeps my body soluble': 'Outwardly I use a decoction of elecampane, dock-root and chalk in whey, twice a day bathing the affected places therewith. Mercury I dare not be bold with.' He admitted that he was not following Sloane's suggestion that he be bled. None of these treatments cured the sores.[50] Four years later, as he wrote to Sloane, another friend had another try:

> Sir Thomas Millington coming to see me, discovering my condition told me that he believed no outward application would do me any good and advised me to use a plain anti-scorbutic diet-drink, made of dock roots, water-cress, brooklime, plantain and alder-leaves which I have done now this fortnight but have yet have received no sensible benefit by it, my sores running as bad and being as painful as ever.[51]

The last years of Ray's life, when he was completing the final volume of his monumental *History of Plants*, were spent entirely confined to his rural home, marooned by his physical ailments.

Ray was one of those who shared the vigil at the dying John Wilkins's bedside, on 18 November 1672. Like Hooke he had been a protégé since the time when Wilkins was briefly Master of Trinity College, Cambridge (where Ray was a fellow for twelve years). During the same years in the 1660s when Hooke and Wilkins were carrying out mechanical experiments together in and around London, Ray and Wilkins were working together on a grand taxonomy of the natural world, its complete classification into fundamental categories, subdivided according to distinguishing features. Working with his close associate Francis Willoughby, Ray provided the systematic coverage of 'plants, quadrupeds, birds and fishes', reducing the entirety of known species to organised groupings linked by shared attributes. On the basis of these tables, Wilkins devised his 'universal language', a code that supposedly connected words scientifically to the things they designated, according to their place in the classification.

Wilkins's *Essay towards a Real Character and Philosophical Language* was a scientific flop – obsolete as a venture almost before it was printed. Its publishing fortunes should perhaps have been treated as a omen. The *Real Character* was in the process of being printed in autumn 1666, when the Great Fire of London in September destroyed the entire print run, which had been stored in a warehouse in St Paul's churchyard. Fortunately Wilkins's own proof copy survived, and he immediately enlisted Ray's and Willoughby's help to redraft and expand with a view to reprinting. They loyally rushed new, enlarged tables out (Ray claimed that the tables of plants were done in three weeks). Wilkins announced optimistically that if the tables could be completed to his satisfaction 'I would put out the next edition in folio with handsome cuts of all such things as are fit to be represented in figure'. The resulting 1668 edition is indeed a handsome volume, with several large, fold-out tables, and a woodcut of Noah's Ark (the archetypal example of complete enumeration of species). But Wilkins's 'methodical enumeration' of the entire natural world is in the end transparently haphazard and incomplete, his 'universal language' a cumbersome code few tried to put into practice.[52]

The reason for this is apparent in Ray's own independent botanical work. He began compiling his global survey, the *History of Plants*, in 1682. In 1684 he began corresponding regularly with Sloane, whom he had recently met and who enthusiastically offered to help with the *History*. Sloane loaned books and specimens to Ray (he also sent the impoverished Ray and his family regular gifts of sweetmeats and venison).[53] He supplied lists of rarities in his own gardens and in the Chelsea Physic Garden, he sent seeds for Ray to grow, and he urged the inclusion of plants of particular interest to himself. He forwarded to Ray any inquiries from the Royal Society concerning botanical matters – for instance, a request to examine a specimen of Peruvian bark to see whether it was, in fact, a fraudulent product, made by dipping black cherry bark in aloes.[54]

Thus, when the first volume of the *History of Plants* came out (with no illustrations) in 1686, it was already significantly marked by Sloane's distinctive interests, and his access to materials. When Ray dealt with plants from other continents, many of which were by this time growing in the great gardens of Europe, it is apparent that he has had to rely on existing publications, supplemented by dried specimens provided by his well-connected correspondents. As a

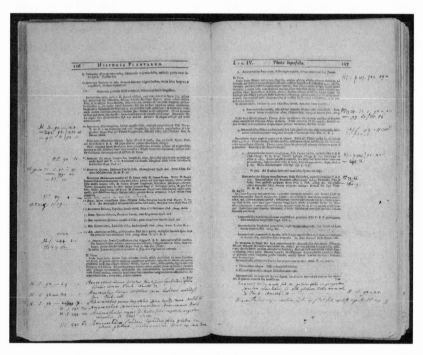

9. Annotated copy of Ray's *History of Plants* used as a partial index to
Sloane's herbarium

result, these make a comparatively poor showing, tacked on to much fuller descriptions of native English plants. Ray himself admitted that 'dried specimens cannot represent all the principal parts, flower, seed-vessel or fruit, and seed: and I have seen none of late discovery growing in gardens, not having ability to take journeys to visit them'.[55]

By the time the final volume of the *History* was published in 1704, all the first-hand observation and material from Sloane's voyage to Jamaica had been incorporated (a copy of the intervening second volume was shipped to Sloane in Jamaica in April 1688). Ray wrote in his preface: 'Now in our day botanists fully trained not only in the European flora but in the contents of public and private gardens have themselves travelled to America, Africa, the East Indies, Japan and China and have studied, collected and described the indigenous plants, thus adding an incredible number of new species.'[56] The fact is, however, that it was through wealthy patrons like Sloane (and, as ever, Petiver, who also corresponded with and supplied Ray) that such materials were made accessible to the taxonomist confined to his home in rural England. Their interests, therefore, inevitably influenced the coverage and where the botanical emphasis was laid in a work like the *History of Plants*.

Nowhere is this clearer than in Ray's later treatment of the flora of North America. For this Ray relied almost entirely on dried specimens and written descriptions sent back from the English colonies, by way of patrons like Sloane and Petiver. In 1692, John Banister, who had been sending plant specimens home from Virginia, was killed in a rock-climbing accident. His correspondents in England, for whom North America was of particular interest botanically as a potential source of marketable medicinal plant materials, hurriedly cast around for a replacement. Edward Lhwyd, Keeper of the Ashmolean Museum at the University of Oxford, came up with a potential candidate: a young Welsh student working there, Hugh Jones, who was willing and able, although he

10. Specimen of cocoa plant from Sloane's collection

had 'no skill at all in plants'. After a few months of intensive training, and a rushed ordination by the Bishop of London, Henry
Compton, Hugh Jones was sent to Maryland to collect plants and
serve as minister to Christ Church Parish in Calvert County. In
1698 the Royal Society sponsored another young botanist, William

Vernon, to join him, while James Petiver sent his own choice, David Krieg.[57]

These three collected multiple specimens of new plants they found, made them up into sets, and sent them home to each of the collectors who had invested financially in the venture. Thus both Petiver and Sloane received identical sets of the more than six

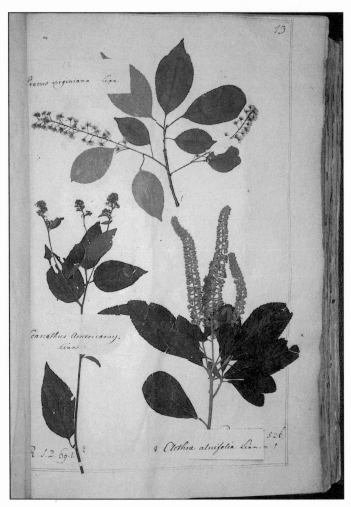

11. Sheet of Maryland plants from Sloane's collection

hundred different species and two thousand specimens they gathered. Sloane's specimens were immediately sent to Ray, who
complained bitterly about their condition, especially that they had
been collected without their roots. Nevertheless, he annotated them
(his labels survive in the Sloane herbarium), and included them in
his own global survey, thereby adding some two hundred species he
might otherwise never have encountered.[58] In the years leading up to
his death in 1705, the bulk of Ray's new material came from his
London benefactors, with Sloane dominating in the field of plants,
while Petiver exercised an equally strong influence on his classification of insects (in which he took a special interest).

When Francis Willoughby died unexpectedly at the age of thirty-
seven, Ray took over publication of his friend and collaborator's
related taxonomic projects – a *History of Birds* and a *History of Fishes* –
working on these alongside his own botanical works. The glory of
the second of these Willoughby volumes (into both of which Ray
put an enormous amount of effort, as a lasting memorial to his
friend) is undoubtedly its illustrations. Financing them proved a difficult matter. The plates were far and away the most expensive part
of any natural history publication – Ray could not find a backer for
any plates at all for the first volume of his *History of Plants*, and
patrons like Petiver made substantial contributions towards the
plates in later editions. Willoughby's widow had paid for the modest
plates in the *History of Birds*. When it came to the *History of Fishes*, it
was, in the end, the hard-pressed Royal Society that agreed to foot
the bill – a particular irony, since they refused to pay for plates for
the first volume of Ray's own *History of Plants*, published in the
same year.

Ray passed the manuscript of the *History of Fishes* to the Royal
Society in March 1685. Pepys, who was now President, was enthusiastic. Finding a printer, however, proved unexpectedly difficult.
Martin, the Society's printer for many years, had recently died, and
when the University Press in Oxford was approached, the Bishop of

Oxford was not keen to take on what was obviously going to be an expensive job without subscriptions. He insisted that the Society must not only finance the entire publication but additionally agree to take one hundred copies of it themselves. More trouble ensued when Ray revealed that many of the plates were to be copied from existing publications, rather than executed especially for the Willoughby volume. Finally William Faithorne's London press agreed to print the illustrations separately (Faithorne took a special interest in engraving), and as a result got the job as the Royal Society's regular printer to replace Martin.[59]

Several familiar figures in the Royal Society circle helped bring the *History of Fishes* to completion. Martin Lister chaired the meeting at which it was finally agreed that the Society would foot the bill for publication costs, and personally took on the job of organising the plates – he persuaded William Courten, among others, to lend specimens from his collection for the drawings. Pepys contributed fifty pounds (thereby paying for seventy-nine illustrations); Wren, Boyle, Evelyn, Ashmole and Sloane all made more modest contributions.[60] The two-volume work, when it finally appeared, was a sumptuous production, which somewhat upstaged Ray's own, much more modest publication, issued two months later.

Nonetheless, the Bishop of Oxford turned out to have been right. The times were politically uncertain, and fishes, in any case, perhaps did not have quite the exotic visual appeal of birds and plants, then as now. The Society had worked out carefully how to cover expenditure with the sale of their hundred copies: members of the Society and those who had contributed plates were to pay £1 for a copy on cheaper paper, and £1. 3s. 0d. for a copy on the best paper, others would be charged £1. 5s. 0d. and £1. 8s. 0d. Although at these prices the Society would barely break even, they were too high for the general public. Hooke was unsuccessful in his approaches to an Amsterdam bookseller, in the hope of disposing of a number of copies abroad.[61]

12 & 13. Plates from Willoughby's *History of Fishes*: perch and gurnard

Thus it was that in July 1686, when Edmond Halley (as one of his first projects as salaried clerk to the Royal Society) proposed replicating Picard's method of finding the arc of the meridian by triangulation, the derisory fee the Society was prepared to offer him for the task was fifty pounds – or fifty copies of the *History of Fishes*.[62] Thus it was also that a year later Halley was obliged to accept his entire salary as clerk to the Royal Society in kind, in the form of copies of the *History of Fishes*, at the members' price of £1 apiece (fifty copies, again).[63]

A more far-reaching consequence of the Society's imprudent investment in Willoughby's and Ray's *History of Fishes* was also closely associated with Edmond Halley. It was in 1687 that Halley finally brought the completed manuscript of Newton's *Principia* to the Royal Society, a work they had agreed to license for publication the previous year. But by 1687 it was clear to the Council of the Society that their previous publishing venture had been a financial catastrophe: they were left with large debts outstanding and unwanted piles of the *History of Fishes*. It was decided that the Society could put no money whatsoever towards the publication of Newton's *magnum opus*. Instead Halley, who had made a personal commitment to Newton that the *Principia* would be published at no expense to himself, and who had fortunately become extremely prosperous on the proceeds of his exploits in his 'other job' as sea-captain, financed publication personally.

Thus Halley became the entrepreneurial owner of Newton's printed *Principia*, with a serious financial interest in seeing it sell. In July 1687 he wrote to Newton to tell him that the book was finally printed and ready for distribution:

> I have at length brought your Book to an end, and hope it will please you. The last errata came just in time to be inserted. I will present from you the books you desire to the R. Society, Mr Boyle, Mr Pagit, Mr Flamsteed and if there be any else in

town that you design to gratifie that way; and I have sent you to bestow on your friends in the University 20 Copies, which I entreat you to accept.

Newton was going to have to do his bit, however, to see that Halley's investment was not a total write-off:

> In the same parcell you will receive 40 more copies, which, having no acquaintance in Cambridge, I must entreat you to put into the hands of one or more of your ablest Booksellers to dispose of them: I intend the price of them bound in Calves leather and lettered to be 9 shillings here, those I send you I value in Quires at 6: shillings to take my money as they are sold, or at 5. shillings a price certain for ready or else at some short time; for I am satisfied that there is no dealing in books without interesting the Booksellers, and I am contented to lett them go halves with me, rather than have your excellent Work smothered by their combinations.[64]

When Halley became clerk of the Royal Society, his unsuccessful competitor for the office was Hans Sloane. Eight years later, Sloane, now much travelled and wealthy, was elected second secretary, becoming first secretary the following year. He took over the publication of the *Philosophical Transactions*, which had lapsed through lack of anyone to administer them, and revived the vigorous secretarial correspondence that had had its heyday under Oldenburg. Under his influence the scientific attention of the Society was turned firmly towards medicine and natural history and away from mechanical experiments and philosophical inquiry (in which he had neither interest nor aptitude). When Newton died in 1727, Sir Hans Sloane succeeded him as President of the Royal Society, remaining in that office for fourteen years until, at the age of eighty-one, he stepped down because of poor health.

All of which should give us pause for thought, as we cast our

14. Engraving of Newton and Sloane together with the inscription,
'Behold the glories of mathematics and of physic'

minds back to the origins of modern science and the founding
moment of the scientific revolution. Had anyone in the circle close
to the Royal Society been asked, in the years before 1700, where the
future of science lay, they would almost certainly have identified as
its focus the compilations of data and systematic taxonomies that
absorbed the interest of Sloane and his fellow physicians and natural
historians. Even during the period of Newton's presidency, when he
was generally too busy to attend and left the running of the Society
in Sloane's hands, the physics, astronomy and mathematics we associate with the birth of modern science today was a minor, specialist
interest, regarded with a certain distaste because of the personal
quarrels about priority and intellectual ownership these more
abstract domains of inquiry seemed frequently to provoke.

CHAPTER 8

COMMITTED TO PAPER

Hide and seek

IT IS NOT hard to see why in the second half of the seventeenth century Bishop John Wilkins – himself the meeting point in a network of scientific enthusiasts – should have thought that the moment had come to promote a universal language. From Oldenburg's multilingual, Europe-wide correspondence on behalf of the Royal Society, to the published diaries and memoranda of voyages eastwards and westwards, information-sharing seemed the order of the day. These exchanges revealed just how much scientific practitioners had in common, regardless of differences of race, conviction and creed. If, according to Sir Francis Bacon's vision of scientific co-operation, 'many shall pass to and fro, and knowledge will increase', how better to achieve that than by agreeing a spoken and written language of intellectual pursuit, common to all?

Correspondence, copious record-keeping and publication in the Society's journal were formalised within the proceedings of the Royal Society as the defining characteristics of the institution. Even today that systematic compilation of 'work in progress' is a rich source of insight into the development of scientific understanding within the international scientific community.

In 1674, a year after he began sending the Royal Society (via Oldenburg) his observations, amateur microscopist Antoni van Leeuwenhoek suggested in his weekly letter that the mechanism of plant growth could be understood by examining cross-sections of plant material under the microscope:

> Which kind of progress of growing I apprehend may in some manner be seen in the pith of Wood, in Cork, the pith of the Elder, as also in the white of a Quill, of which three last I have sent you and your curious Friends some small particles, cut off with a sharp razor.[1]

Van Leeuwenhoek's choice of specimens to examine here confirms that by 1674 he had been able to consult the text of Hooke's *Micrographia*. John Wilkins had undertaken a Latin translation in 1665 (while he and Hooke were at Durdans), but in spite of pressure from Oldenburg, it was never completed.[2] So someone in Delft must have translated Hooke's text into Dutch for van Leeuwenhoek. Alongside his illustration of the minute structure of a section of cork, Hooke proposed that the honeycomb-like structure seen revealed plant life to be composed of 'cells' (the first recorded occasion on which this vital suggestion was made). He went on:

> Nor is this kind of Texture peculiar to Cork onely; for upon examination with my Microscope, I have found that the pith of an Elder, or almost any other Tree, have much the same kind of Schematism, as I have lately shown that of Cork. The pith that also fills that part of the stalk of a feather that is above the Quil has much such a kind of texture.[3]

Van Leeuwenhoek chose to examine the identical sequence of specimens to those suggested in Hooke's text, and now sent prepared

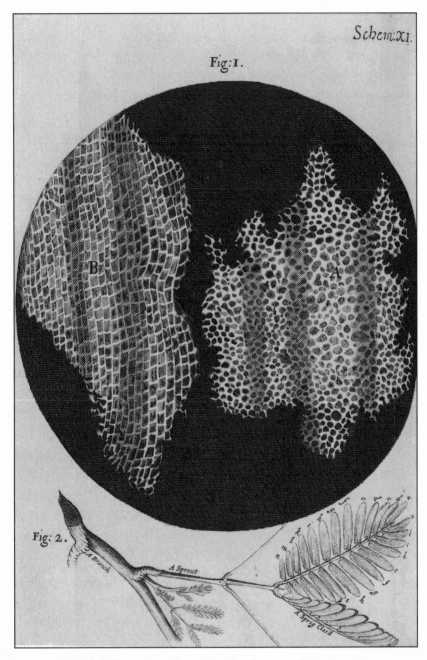

1. Hooke's illustration of a section of cork seen under a microscope

specimens so that the members of the Royal Society could examine both Hooke's claim and his own cross-checking of it. Since the extraordinary results van Leeuwenhoek got with his microscope were due in large part to his practical expertise in preparing his materials, this was no small help to on-going collaborative researches in microscopy.

In this case there is a further stage in the careful exchange of scientific information, which reveals rather dramatically the importance of such exchanges being meticulously recorded and filed for posterity by the Royal Society. In 1981, a modern micro-scopist discovered the very specimens van Leeuwenhoek refers to in his letter in the Royal Society archives, still pasted in their paper envelopes to the end of its final page. He was able to exam-ine these under a modern reconstruction of Leeuwenhoek's single-lens microscope, and so prove how remarkable the magni-fication of that microscope was in the hands of a proficient user. The experiment that had led to the discovery of living cells was reproduced, more than three hundred years after it was first undertaken.[4]

At the heart of the Royal Society's institutionalised procedures lay the 'witnessing' or 'attesting' to experimental outcomes. This was established early on as an absolute requirement for validating results (it remains a feature of the process of putting scientific experimental findings on the record even today). Confirmation of any remarkable finding by a select group of scientifically qualified members was required before an experimental outcome could be officially recorded as fact in the *Philosophical Transactions*. It had also to be shown that the result could be reproduced on more than one occasion. Thus when van Leeuwenhoek reported that he had found 'little animals' in pepper-water, it was twelve months before Hooke succeeded in replicating the experiment – at which point it was performed before senior members of the Royal Society, who could at last record that there was 'no longer any doubt' of

2. Packets of prepared specimens sent by Leeuwenhoek to the Royal
Society

the existence of protozoa: 'They were seen by Sir Christopher
Wren, Sir Jonas Moore, Dr Grew, Mr Aubrey, and divers others;
so that there was no longer any doubt of Mr Leeuwenhoek's
discovery.'[5]

Nevertheless, the idealised picture of information globally
available to all as the seventeenth century's model for the successful transmission and growth of knowledge needs to be
qualified. Reproducibility of experiments was a fundamental cri-
terion for experimental proof. But it was equally widely taken for
granted by the 'virtuosi' that those enthusiasts with mere 'ordinary' curiosity had in the end to be excluded from full
participation. In line with a much older tradition that 'knowledge
is power', arcane learning was to remain the preserve of those who
had proved themselves of sufficient intellectual ability, and who
had earned the trust of the specialist scientific community. Out

of the hurly-burly of international scientific dialogue, an élite of those who had shown themselves worthy to participate would gradually emerge.[6]

Denying access to vital knowledge might, in any case, play a key part in a shaping a scientific discovery, as in the case of the clash between Hooke and Huygens over who had invented the balance-spring regulator for a pocket-watch. To prove the superiority of his watch, Hooke put it out to trial with several prominent London figures, sealing the case of the watch so that no one could examine the mechanism. To claim a patent for his invention, however, he needed to put the detail of his device on paper, and subject it to expert scrutiny.[7]

One established route to such scrutiny was to publish the details

3. Teniers's vivid image of an alchemist (or early chemist)

in the *Philosophical Transactions* of the Royal Society, a journal with a suitably restricted circulation among scientific initiates. But Oldenburg was reluctant to accept Hooke's manuscript for publication since he had officially backed Huygens's counter-claim on behalf of the Society (and had already reprinted the equivalent account of Huygens's watch mechanism, originally published in the equivalent journal of the French Académie). When Hooke became convinced that Oldenburg was actively trying to suppress his claim, he took the unusual step of adding an account of his invention as a postscript to his own *Description of Helioscopes*, which was on the point of publication. After the body of the text (on an entirely different subject) Hooke inserted an encrypted description 'of the general ground of my Invention for Pocket-Watches' together with pointed criticism of Oldenburg's conduct in the matter.

Hooke collected the first copies of his book from the printer on 11 October 1675. The dispute between himself and Oldenburg was aggravated by the fact that Hooke's printer, John Martin, was the officially designated printer to the Royal Society. When Oldenburg came to Martin's print shop on 9 October, Hooke was there proof-correcting the final sheet of added material. 'Corrected last sheet at Martins. Oldenburg there. Said I should have appeald to Councell,' Hooke recorded in his diary.[8] Two days later, in a state of great agitation, Oldenburg wrote to Huygens, asking him to write to the President of the Royal Society, rebutting Hooke's printed accusation that 'I am your spy here for communicating of everything of note found out here, and that I wanted to defraud him of the profit of his invention'.[9] Martin, the printer, found himself caught in the middle. At one point Oldenburg tried to have him deprived of his entitlement to print Royal Society publications because he had connived in Hooke's additions to his manuscripts. The matter was dropped when Martin insisted he had had no knowledge of the nature of the added material.

The whole affair demonstrates the trickiness of at once encouraging publication as evidence of priority and intellectual property ownership, and at the same time having key individuals, like Oldenburg, in positions of power over such publications. Hooke maintained that Oldenburg had deliberately failed to print important inventions of his, while passing the information in letters to other inventors like Huygens. Oldenburg was certainly vulnerable to such charges, and senior Royal Society members like Boyle advised that careful procedures of dating correspondence and material received should be adopted to protect the Society.

The code Hooke chose to record the secret of his balance-spring watch in his *Description of Helioscopes* was none other than Wilkins's 'real character'. He explained to his readers that the idea was that the code should serve as a kind of initiation test:

> In order that the true Lovers of Art, and they only may have
> benefit of it, I have set it down in the Universal and Real
> Character of the late Reverend Prelate, my Honoured Friend
> Dr. John Wilkins, Lord Bishop of Chester, deceased. In which
> I could wish, that all things of this nature were communicated,
> it being a Character and Language so truly Philosophical, and
> so perfectly and thoroughly Methodical.[10]

There is considerable irony in Hooke's adoption as a code of a language designed to bring clarity and transparency to scientific communication. Wilkins's earlier work, *Mercury, or the Secret and Swift Messenger* (1641), was explicitly a handbook on cryptography, and it appears that Hooke was not alone in regarding the 'real character' as better suited for encoding. Sir Hans Sloane's copy of Wilkins's *Essay towards a Real Character and Philosophical Language*, which survives in the British Library, still contains, interleaved, a sheet of paper with the key to another complicated piece of cryptography.[11] Indeed, although it is a relatively simple matter to turn any sentence into Wilkins's language (using the handy fold-out tables of characters

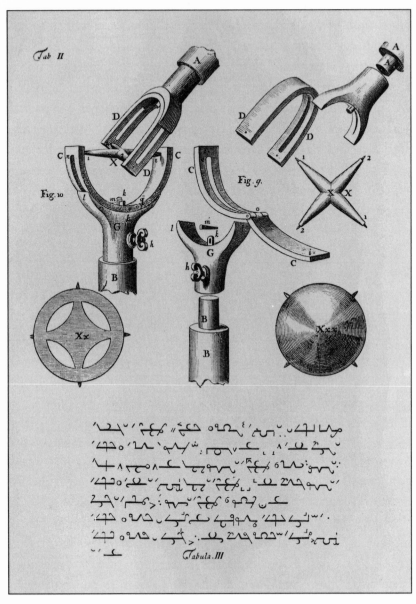

4. Hooke's announcement of his balance-spring watch mechanism,
encrypted in Wilkins's Universal Language

included), it is virtually impossible for another person to reverse the process and return a passage to ordinary language. Hooke's seven-line message remains, as far as I know, untranslated down to the present day (after several hours' trying I managed the first line only). Hooke's closing appeal to his 'Reader who will be at the pains to decypher and understand this description' establishes a personal bond between himself and the person who has succeeded in cracking a code that has made serious demands upon his intelligence and application: 'I have this further to desire, that he would only make use of it for his own information, and not communicate the explication thereof to any that hath not had the same curiosity with himself.'

Hooke might even have meant this last remark facetiously. The point of coded passages like this one of his was not, in fact, primarily to give access to the contents: it was to reserve the author's right to claim later that he had been the first with the knowledge enshrined secretly within it. Coded announcements of discoveries in science were a way of 'place-keeping' for ideas that would be divulged later when they had been fully worked out. The convention applied particularly to ideas put up for patenting, or royal licence. The individual would file for a patent, including with the application a sealed envelope containing a coded explanation of the process. He then went off and ploughed as much funding as he could into resolving the technical details of the invention, returning to claim the patent proper within a period specified in the provisional patent document.

Alongside the coded passage about Hooke's pocket-watch is another (this time a simple anagram), in which Hooke records 'the True theory of Elasticity or Springiness, and a particular Explication thereof in several Subjects in which it is to be found: And the way of computing the velocity of Bodies moved by them. ceiiinosssttuu'.[12] Two years later, in 1677, Hooke went public with what we now know as Hooke's law of elasticity, revealing to his readers the solution to his anagram:

About two years since I printed this Theory in an Anagram at the end of my Book of the Descriptions of Helioscopes, viz. ceiiinosssttuu, that is, *Ut tensio sic vis*; That is, The Power of any Spring is in the same proportion with the Tension thereof: That is, if one power stretch or bend it one space, two will bend it two, and three will bend it three, and so forward. Now as the Theory is very short, so the way of trying it is very easie.[13]

By this time, Hooke was able to provide experimental evidence of his law of elasticity, and to explain the importance of the law for the design of his balance-spring watch, which he naturally chose as one of his worked examples.

From the very start, scientific 'intelligence' (the latest information in scientific matters) was not freely available to all the members of the Royal Society equally, let alone to all and sundry. An early letter from Henry Oldenburg to Robert Boyle, written in 1657, before the Restoration, offers us a delightful example of the way that in the burgeoning scientific community careful reticence about 'restricted' material coexisted with an exuberant desire to communicate and share knowledge. Oldenburg was on the continent with Boyle's nephew, whom he was tutoring. The letter describes how he had met an Italian mathematician who 'had a way of writing to others very secretly'. The Italian was not prepared to exchange the secret for one of Oldenburg's (the established way of acquiring arcane material), but did in the end accept money to reveal it. A colourless solution of lead acetate (made from lead oxide mixed with vinegar) was to be used to write the concealed message. Over this another, innocuous message would be written in carbon ink. The recipient would wash the letter with a solution of arsenic trisulphide and lime-water, which would reduce the carbon and remove the surface message, while at the same time converting the lead acetate into grey lead sulphide, so making the hidden message visible: 'This water will be wonderfully effective in effacing the letters written in ink and in

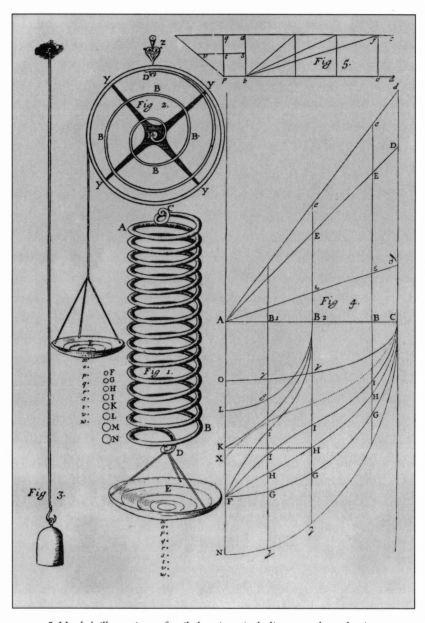

5. Hooke's illustrations of coiled-springs, including a watch mechanism

rendering visible to your friend everything which you have written in white letters.'

(Of course, the chemicals in the 'recipe' as Oldenburg gives it sound far more exotic: 'Distilled vinegar', mixed with 'Litharge in the form of stone'; powdered charcoal to which is added 'a little mucilage, or gum, to make an ink'; 'I ounce of orpiment' pulverized and mixed with '2 ounces of quicklime, good and strong', plus ordinary water, for the washing solution.)[14]

Bartering secret information played an important part in Oldenburg's vast and apparently generously open correspondence. In 1659 Oldenburg, for instance, wrote to Samuel Hartlib, whose scientific circle in Oxford included Boyle, and prefigured the Royal Society: 'I send you here enclosed a Chymical process of vitrioll (in acknowledgement of the secret you sent me, which shall not loose the name of a secret for me). And intreat you, to communicate it to none, but noble Mr. Boyle.' Martin Lister wrote to Oldenburg in 1672: 'I am willing to entertain you with my thoughts upon the analysis of mineral waters; but desire nothing of this nature from me may be made publick by the presse for quiet sake.'[15] Chemical and alchemical processes were evidently regarded as sensitive items, and the reproducibility of experiments in these fields was narrowly restricted to small groups of individuals who trusted one another. Such is the emphasis on open exchange of knowledge we understand to have operated within the Royal Society that it comes as something of a surprise to find figures of the scientific standing of Boyle and Newton corresponding surreptitiously, trading secrets.

Guarding secrets

Robert Boyle died aged sixty-four, on New Year's Eve 1691. Early in 1692, Newton, as nonchalantly as he could, added a postscript to a letter addressed to Locke, who was one of the close friends Boyle had appointed to sort his papers, asking about an alchemical process for which he believed Boyle held instructions. 'I understand,' Newton wrote, 'that Mr Boyle communicated his process about the red earth and mercury to you as well as to me, & before his death procured some of that earth for his friends.'[16] Locke responded by sending Newton some of the 'red earth' that he was after – more, indeed, than Newton had really wanted. He was only curious, he insisted, and had no intention of trying the process himself:

> For I desired only a specimen, having no inclination to
> prosecute the process. For in good earnest I have no opinion of
> it. But since you have a mind to prosecute it I should be glad to
> assist you all I can, having a liberty of communication allowed
> me by Mr B. in one case which reaches to you if it be done
> under the same conditions in which I stand obliged to Mr B.
> For I presume you are already under the same obligations to
> him. But I feare I have lost the first & third part out of my
> pocket.[17]

In August, however, Newton took up the matter of the tripartite process with Locke again. He was sure that the first stage produced the 'philosophical mercury' about which Boyle had written in the *Philosophical Transactions* over fifteen years earlier, though he believed that Boyle himself had never tried it. Despite the affected casualness, Newton's eagerness to obtain the full 'receipt' for the secret process is transparent:

> But besides if I would try this recipe, I am satisfied that I could
> not. For Mr B. has reserved a part of it from my knowledge. I
> know more of it than he has told me, & by that & an
> expression or two which dropt from him I know that what he
> has told me is imperfect and useless without knowing more
> then I do. And therefore I intend only to try whether I know
> enough to make a mercury which will grow hot with gold
> [incalescent mercury], if perhaps I shall try that.

Finally, Newton came to the point and asked Locke directly for the instructions to the first part of the process (which one imagines was supposed to lead, in its third stage, to the elusive 'elixir of life'). He would make do with the recipe for the mercury, which becomes warm to the hand when correctly mixed with a small amount of gold:

> Pray will you be so free as to let me know whether he communicated to you any thing more than is written down in the 3 parts of the recipe. I do not desire to know what he has communicated but rather that you would keep the particulars from me (at least in the 2^d & 3^d part of the recipe) because I have no mind to be concerned with this recipe any further then just to know the entrance.[18]

Knowing the second and third stages in the secret process was just too much of a responsibility, according to Newton. Boyle had offered them to him as part of a mutual exchange of restricted information. He had given Boyle instructions for 'a certain experiment' of his own, in exchange for which Boyle had sent instructions for two others. Boyle had made it a condition of the transaction, however, that Newton should carry out the experiments contained in the 'recipes', and publish the results after Boyle's death. Newton had done nothing of the sort, and now requested that Locke include them (untested) with any others he intended to publish – without, however, associating Newton's name with them in any way. At the end of this self-conscious set of exchanges, Locke transcribed and sent to Newton a peculiar process on mercury, written in code.[19]

The process Newton was so eager to get his hands on was a chemical one to produce a mercury compound that, when mixed with a small amount of gold, speedily absorbed it, and became hot to the hand. This 'philosophical' or 'incalescent' mercury was regarded as indispensable for the preparation of the philosophers' stone (the goal and end-point of all alchemical pursuit), as a solvent

able to dissolve gold 'radically'. Boyle had publicly acknowledged his
interest in the incalescent mercury in 1676, when he published a
paper in the *Philosophical Transactions* of the Royal Society entitled 'Of
the Incalescence of Quicksilver with Gold', using the thinly dis-
guised alias 'B. R.', which no one at the time seems to have had the
slightest difficulty in deciphering.[20]

In his paper Boyle claimed that he had successfully made such a
philosophical mercury in his own laboratory. By distillation,
common mercury could be freed from heterogeneous 'recrementi-
tious' parts, and made to incorporate other materials to render its
effects (though not its outward appearance) quite different from
normal. When he mixed this substance with finely powdered gold,
in the palm of his hand, he continued, he felt a distinct and sometimes strong heat. To verify his claim, the procedure had been
performed in the hands of Oldenburg and Brouncker, both of
whom confirmed the heating effect.

Boyle's account was extremely carefully judged. While he
described in detail the warming effect of the mixing of the incalescent mercury with powdered gold, he never mentioned the method
by which normal mercury was purified, nor what the cleansing
agents were. The secrecy surrounding the process was heightened by
a brief preface to the paper provided by Oldenburg, in which he
requested that the reader 'not fruitlessly endeavour to put the
author upon making unseasonable answers to any Verbal or
Epistolary Questions about things, wherein some considerations,
that he thinks are not to be dispensed with by him, do as yet injoyn
him silence'.[21] The author was not at liberty to divulge his process,
or as he himself wrote to Oldenburg, 'Divers Queries and perhaps
Requests (relating to this Mercury) I would by all means avoid, for
divers reasons.'[22]

Although Boyle's paper is presented with all the scientific apparatus typical of proceedings at the Royal Society, his text makes it
clear that Boyle believes the manufacture of incalescent mercury to

be one stage along the route to traditional alchemy's dream of the philosophers' stone and the transmutation of base metal into gold. The details of the method of manufacture of the mercury have to be kept secret because, like all alchemical secrets, only a highly select band of initiates is allowed access for fear of the process falling into the wrong hands.

If the process was such a secret one, why did Boyle publish it in the *Philosophical Transactions*? In 1689 Newton told his young colleague Fatio de Duillier, who was also closely interested in alchemical processes, that he had stopped corresponding with Boyle on such matters because he conversed with all sorts of people and was 'in my opinion too open & too desirous of fame'.[23] But it was not fame that Boyle was after here. It seems that Boyle was unable to find the means of progressing from the philosophical mercury to the philosophers' stone. His article was a carefully judged advertisement to others in the field (within the respectably serious readership of the *Philosophical Transactions*), to come forward with assistance:

> I hope I may safely learn what those that are skilful and
> Judicious enough to deserve to be much considered in such an
> affair, will think of our Mercury. The knowledge of the
> opinions of the wise and skilful about this case will be requisite
> to assist me to take right measures in an affair of this nature.
> And till I receive this information, I am obliged to silence.[24]

So that the call for further information might reach as wide as possible an erudite readership of 'adepts' or alchemical specialists, Boyle arranged for Oldenburg to translate his paper into Latin. It was published in the journal in parallel columns in the two languages – the only occasion in the history of the *Philosophical Transactions* on which this happened.

We do not know what kind of response Boyle got to his announcement, nor whether any 'adepts' came forward to assist him. Such a method of broadcasting invitations for help clearly had its drawbacks. In summer 1677, Boyle was approached by Georges Pierre des Clozets, who claimed to be the agent of the 'Patriarch of Antioch', leader of a society of alchemical masters, whose number Boyle had supposedly been nominated to join. Pierre wrote to Boyle before their meeting to tell him that it would not be long 'before God allows him to enjoy the happiness of being a true philosopher [that is, master of the philosophers' stone]'.[25] The wording suggests that Pierre was responding to Boyle's advertised desire to learn the alchemical processes following on from the manufacture of incalescent mercury. Whatever chemical materials or processes he showed to Boyle, they must have been impressive. Over the next twelve months, Boyle bombarded Pierre with expensive gifts for the 'Patriarch' to guarantee his admission to the society: a large telescope, assay balances, a globe, copies of the New Testament in Turkish, and a case of one hundred glass vials for Pierre's laboratory. He sent gifts for the Turkish court where the 'Patriarch' was supposed to reside: jackets, fine fabric ('for the sultana queen mother, eight rods of flesh-coloured moiré, eight of gold-coloured moiré, and eight of flame-coloured moiré'), and a chiming clock ('more than three feet high').[26]

Boyle was never admitted to the Patriarch of Antioch's secret society, and the whole affair appears to have been an elaborate hoax, in the course of which the participants extracted large sums in money and lavish gifts from a gullible Boyle, who got little or nothing in return.[27]

A rather more respectable respondent to Boyle's paper was Newton himself. On 26 April 1676 he wrote to Oldenburg, correctly identifying Boyle as the article's author, and commending his secrecy over the details of his method of making incalescent mercury: 'he does prudently in being reserved'. 'The great wisdom of the noble author will sway him to silence,' Newton went on. For if this was in fact the true philosophical mercury, it could not be communicated generally 'without immense dammage to the world' (global political upheaval was the usual reason given for alchemists' reticence on their methods for the unlimited production of gold). However, Newton confided to Oldenburg that he doubted if Boyle had actually carried out the process successfully.

Two weeks on, Newton was far less sceptical. We may assume that in the meantime he had corresponded with Boyle, or been shown the process by him, or both. In a second letter to Oldenburg Newton simply stated: 'I perceive I went upon a wrong supposition in what I wrote concerning Mr. Boyle's experiment.' The fact that Newton's letters to Locke after Boyle's death refer to regular alchemical correspondence with Boyle makes it clear that this was only the beginning of Newton's contacts with Boyle on alchemical matters. And when Locke finally sent him Boyle's recipe for the incalescent mercury, Newton, as one of the alchemical elect, had no difficulty in translating it out of the protective code in which it was written. We do not know if he tried the process himself. But it was later that same year that Newton suffered his well-known mental breakdown, which might plausibly have been brought on by acute mercury poisoning – an overdose of mercury in the system, the result of performing repeated distillations and digestions of mercury in the confined space of his tiny laboratory at Trinity (modern microanalysis of samples of Newton's hair has revealed it to contain very high levels of mercury).[28]

Curved by nature

There was one area of new knowledge that had no need of sophisticated ciphers in order to restrict access to initiates. The language of pure mathematics was at once international and totally obscure to all but a small number of talented individuals who had mastered its conventions and techniques. In the course of the seventeenth century mathematics developed alongside the advances in technology at a remarkable speed as the second essential component for the solution of problems of the movements of heavenly bodies, in navigation and in cartography. France led the rest of Europe in pure mathematics throughout the seventeenth century.

In 1693 the mathematician Gottfried Wilhelm Leibniz wrote to Huygens – who, so far as mathematics was concerned, counted as a Frenchman, since his mature work was done as a leading member of the French Académie: 'You greatly honour Geometry, when you find the most beautiful applications for the curves she provides.'[29]

It was a shrewd comment. Since adolescence Huygens had been fascinated by the mathematical properties of naturally occurring curved surfaces. The central portion of Huygens's definitive work on pendulum clocks, the *Horologium Oscillatorium*, published in 1673, is a masterly mathematical analysis of the cycloid, the curve traced by a point on the rim of a wheel as it rolls on a horizontal surface.[30] Huygens's interest in the cycloid was straightforward: the mathematical properties of the curve meant that it ensured the perfectly regular swing of the pendulum of his clock. When asked why he did not pay more attention to more obscure, related curves, Huygens retorted that he had no interest in curves that did not occur frequently in the natural world.

But the curve that intrigued Huygens throughout his life was the catenary, the U-shaped loop of a simple, heavy chain hanging under gravity. Huygens's very first piece of mathematical work, sent by his proud father to Mersenne for comment, proved that the hanging-chain curve was not a parabola as Galileo had believed. Much later in his life, in 1690 Jacob Bernoulli issued a challenge to all comers (all brilliant mathematical comers), advertised internationally in the journal of pure mathematicians, the French *Acta eruditorum*, to find the curve of a hanging chain. Huygens pointed out to Leibniz that he had dealt with the catenary 'when I was only fifteen' (actually he was seventeen). As a result of the competition, the correct equation for the hanging chain was finally arrived at and published in the *Acta eruditorum* in 1691. Huygens was indeed one of those who came up with the correct solution.

The definitive form of the dome of St Paul's Cathedral in London was settled some time after 1702.[31] Robert Hooke's library contains copies of the relevant issues of the *Acta eruditorum*, so we know he was abreast of developments in the mathematics of hanging chains. The inverted catenary-shaped masonry that supports the dome of St Paul's depended on a problem in mathematics solved only a year or so previously.

Of course, Hooke had proposed his 'true Mathematical and Mechanichal [*sic*] form of all manner of Arches for Building, with the true butment necessary to each of them' many years before, in the early 1670s.[32] But, then, Hooke was never above a bit of bluffing when the theoretical details of an invention eluded him. The minutes of the Royal Society record that, on 7 December 1671, 'Mr. Hooke produced the representation of the figure of the arch of a cupola for the sustaining such and such determinate weights, and found it to be a cubico-paraboloid conoid; adding that by this figure might be determined all the difficulties in architecture about

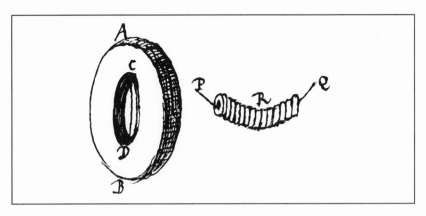

6. Original drawing from Huygens's 1646 paper on the catenary showing idealised hanging chain

arches and butments.'[33] It is possible Hooke got his 'cubico-paraboloid conoid' from an early draft of Huygens's *Horologium oscillatorium*, in which there is a discussion of the semicubical paraboloid, a curve related to the parabola but which does not even slightly resemble the catenary. At any rate, Hooke's remark is pure bluster. Although his hunch was correct about the structural strength of the inverted catenary, neither he nor anyone else in 1671 knew what the equation of that curve was.

Reader, I married him

Those who, for one reason or another, were not entitled to figure in the written records of institutions like the Royal Society or semi-official bodies of scientific correspondence like Oldenburg's have inevitably disappeared. By the end of Boyle's life, the technicians who followed Hooke and Papin into his service were required to sign a document in the form of a solemn oath, which bound them to complete secrecy with regard to their master's scientific work:

> Whereas I _____ being now in ye service of Mr. Boyle he is pleas'd to imploy me about ye making of divers Experiments that he would not have to be divulg'd; I do hereby solemnly and faithfully promise & ingage myself that I wil be true to ye trust repos'd by my sayd master in me, that I will not knowingly discover to any person whatsoever, whether directly or indirectly, any process, medicine, or other Experiment, which he shal injoin me to keep secret & not impart; without his consent first obtain'd to communicate it. And this I promise in ye faith of a Christian, witnes my hand this _____ day of _____.[34]

Boyle had never carried out much in the way of practical work himself, preferring to leave dissections and experiments with dangerous chemical compounds to his assistants. Rarely named (unless like Hooke and Papin they had gone on to make names for themselves with their own independent experimental work),[35] these assistants now disappeared entirely from the record, mute agents of their constitutionally delicate and practically diffident master.[36]

There were other categories of skilled and dedicated experimentalists who suffered a similar fate. In 1712, Maria Winkelmann lost her long battle to become Academy Astronomer at the Academy of Sciences in Berlin. Born in 1670 in Leipzig, Winkelmann showed early aptitude for observational astronomy, and became an unofficial apprentice of the astronomer Christoph Arnold, who lived nearby. She was not, however, allowed to assume any official position in the German astronomical community. Winkelmann took the only route available to her; she married Gottfried Kirsch, thirty years her senior, the leading astronomer in Germany, who had trained with Hevelius in Gdansk. In 1700 the two moved to Berlin, where Kirsch became astronomer at the new Royal Academy of Sciences.

In 1702, Maria Winkelmann was the first to spot a comet in the night sky. Her husband described her discovery:

> Early in the morning (about 2 a.m.) the sky was clear and starry. Some nights before, I had observed a variable star, and my wife (as I slept) wanted to find and see it for herself. In so doing she found a comet in the sky. At which time she woke me, and I found that it was indeed a comet. I was surprised that I had not seen it the night before.[37]

Since this was the first astronomical 'discovery' of the new Academy, a report was sent immediately to the King. It bore Kirsch's, not Winkelmann's name, however, as did all contemporary published accounts, effectively erasing her from the record. (Leibniz, though, described Winkelmann's talents with admiration: 'She observes with the best observers, she knows how to handle the quadrant and the telescope marvellously.')[38]

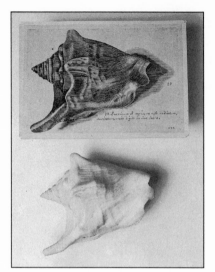

7 & 8. Engravings, with the shells observed, produced by Lister's daughters for his *Historia Conchyliorum*. The shells form part of Sloane's collection

When Kirsch died, in 1710, his widow applied for the vacancy as Academy Astronomer. The secretary of the Academy wrote to Leibniz, who was on the appointments committee: 'You should be aware that this approaching decision could serve as a precedent. We are tentatively of the opinion that this case must be judged not only on its present merits but also as it could serve as an example in the future.'[39]

Maria Winkelmann did not get the job, though Leibniz did, in fact, continue to support her.

Hevelius, in whose private observatory Kirsch had worked as an apprentice, had also married an able young woman astronomer many years his junior (he was fifty-one, she sixteen when they married). In 1679, when Edmond Halley returned to London from his Royal Society-backed visit to Gdansk to observe Hevelius's method of taking astronomical measurements, he ordered an expensive dress, in the height of fashion, for Elizabeth Hevelius: it was made of ten yards of outstanding silk, and the petticoat of eight yards of silk. The total cost, including the tailor's time, was twenty-seven and a half imperiales. Halley wrote to Elizabeth explaining that he had eighteen imperiales in hand from Hevelius (they were supposed to be spent on books he had ordered), so he was owed nine and a half imperiales. If she would send him a copy of each of Hevelius's books (on the moon, and on comets), he would consider the account settled.[40]

Elizabeth Hevelius and Edmond Halley had carried out observations together while he was in Gdansk. They were a mere ten years apart in age, she being the elder. The dress clinched it. When Halley was Savilian Professor at Oxford, the assistant librarian at the Bodleian commented on the fact that Halley's new portrait hung next to Hevelius's in the Library: 'Some persons say that Halley made him a Cuckold, by lying with his wife when he was at Gdansk, the said Hevelius having a very pretty Woman as his Wife.'[41] Such was the fate of the reputation of a good-looking woman with scientific aptitude, then as now. Invisible assistants to distinguished fathers, husbands and brothers, they are made visible only in the sexual gossip that surrounds any unattached man with whom they come in contact.

Book burning

The Dutch perspective painter Samuel van Hoogstraten, resident in England in the 1660s, recorded his reaction to the Great Fire of London of 1666 in poetry and paint. The painting has not survived, but the Dutch poem, 'Fleeting fame, and the pitiful account of the Fire of the City of London' was published in 1669. Van Hoogstraten quoted several lines from it in his treatise on perspective painting, where he described the curious optical effects of smoke, vapour and mist:

Smoke and mist both obscure shadows more than they do light.
But when I am speaking of smoke, I cannot forget that
beautiful yet pitiable morning, the 12th of September, 1666. It
was Sunday, and I was busy working amidst my books in White
Street in London, when I became amazed at how red and
glowing the rays of the sun shone in my room. Whereupon I
went to the window and saw that a pink smoke, which I
mistook for clouds, was blowing toward the southwest.[42]

One of the lasting impressions the Great Fire of 1666 seems to
have left with Londoners caught up in it was the physical vulnerability of books. Because St Paul's was the printing and bookselling
sector of old London, books – in bulk – were a major casualty of
the conflagration, and seem to have been at the forefront of the
minds of those caught up in the catastrophe. Pepys recorded that
the Sunday after the fire he heard a sermon preached in which the
Dean of Rochester said that 'at this time the City is reduced from
a large Folio to a Decimo tertio'. It is striking how many of the
Royal Society circle were preoccupied, after 1666, with building significant collections of books, and guarding them assiduously against
the ravages of time – both in the care with which they were housed,
and the attention given to wills and inventories of personal effects,
in the hope that their libraries might be preserved for posterity.

Two weeks after the fire, on 19 September 1666, Pepys brought
all his 'fine things' back from Bethnal Green and Deptford, where he
had sent them for safety as the fire spread. To his consternation, several of his precious books were missing. He was 'mightily troubled,
and even in my sleep, at my missing four or five of my biggest
books – Speed's Chronicle – and maps, and the two parts of
Waggoner, and a book of Cards'.[43] The following day he was 'much
troubled about my books'. In the evening, as he reshelved and organised his library he was 'mightily troubled for my great books that I
miss. And I am troubled the more, for fear there should be more
missing than what I find.'

Two days later all turned out to be well for Pepys:

W. Hewer tells me that upon enquiry, he doth find that Sir W. Penn hath a hamper more than his own, which he took for a hamper of bottles of wine, and are books in it. I was impatient to see it, but they were carried into a wine-cellar, and the boy is abroad with him at the House, where the Parliament meet today, and the King to be with them. At noon, after dinner, I sent for Harry and he tells me it is so, and brought me by and by my hamper of books, to my great joy, with the same books I missed, and three more great ones and no more – I did give him 5s for his pains; and so home with great joy, and to the setting some of them right.[44]

Not everyone, however, was so lucky. The booksellers in St Paul's churchyard, the heart of the London book trade, lost their entire stock, which they had stored for safety in the Stationers' Hall, St. Faith's Church and Christ Church, all of which burned to the ground. Pepys wrote on 26 September:

Here by Mr. Dugdale [Sir William Dugdale's son] I hear the great loss of books in St. Pauls churchyard, and at their hall also [Stationers' Hall, where newly registered book stock was housed] – which they value at about 150,000 pounds; some booksellers being wholly undone; and among others, they say, my poor Kirton. And Mr. Crumlum, all his books and household stuff burned; they trusting to St. Fayths, and the roof of the church falling, broke the Arch down into the lower church, and so all the goods burned – a very great loss. Sir William Dugdale hath lost about 1000 pounds in books.[45]

In addition to the entire print run of Wilkins's *Essay towards a Real Character and Philosophical Language*,[46] an English version of Galileo's *Dialogue of Two World Systems* (translated from Mersenne's French edition) was apparently totally lost.

9. Contemporary Dutch painting of the Great Fire of London

On 5 October Pepys learned the full extent of his own bookseller Kirton's losses:

> This day, coming home, Mr. Kirton my bookseller's kinsman came in my way; and so I am told by him that Mr. Kirton is utterly undone, and made 2 or 3000 pounds worse than nothing, from being worth 7 or 8000 pounds. That the goods laid in the churchyard fired, through the window, those in St. Fayths church – and those coming to the warehouses doors fired them, and burned all the books and the pillars of the church, so as the roof falling down, broke quite down, which it did not do in the other places of the church, which is alike pillared (which I knew not before); but being not burned, they stand still. All the great booksellers almost undone – not only these, but their warehouses at their hall, and under Christchurch and elsewhere, being all burned. A great want thereof there will be of books, especially Latin books and foreign books; and among others, the polyglottes and new Bibles, which he believes will be presently worth 40 pounds apiece.[47]

The Royal Society's programme of regular publications was directly affected by the fire. In late September Oldenburg wrote to Boyle (who as usual had avoided being caught up in events by staying in Oxford), concerning disruption of the usual arrangements for printing and distribution of the *Transactions*:

> I shall find it very difficult to continue the printing of the Transactions, Martyn and Allestry being undone with the rest of the stationers at Paul's church-yard, and all their books burnt they had carried for safety into St Faith's church, as they call it; besides, that the city lying desolate now, it will be very hard to vend [sell] them at the present.[48]

A month later Oldenburg was still having trouble sorting the matter out. In another letter to Boyle, he made clear how fundamental to the Royal Society's activities he regarded these publications as being. He was even reduced to consider contributing his own labour free of charge:

> The stationers and printers having sustained great losses in the late fire, and not knowing, as yet, how to settle and to reassume their trade, so as to make gain thereby; do very much scruple to print any thing, except it concern the present affair of the war, and of the city: in regard whereof, it will be very difficult to persuade them to continue the printing of the Transactions, unless I let them be printed without consideration for the charges and pains I am at in the digesting of them; as I did the last: which my condition will not bear, however my soul be free enough to consent to it, if I could.[49]

The Fire also had an impact on the running of the Royal Society indirectly. Gresham College, the Society's long-standing head office and regular meeting place (which was also Hooke's permanent home) was requisitioned, under emergency measures, by the City of London, to replace the Guildhall, which had burned to the ground.

Fortunately Henry Howard, Duke of Norfolk, stepped forward and offered the Society the use of his family home, Arundel House on the Strand instead. For a number of years thereafter, most Royal Society activities took place there. Arundel House housed the Arundel Library – a remarkable book and manuscript collection, put together by Howard's grandfather, the distinguished collector of antiquities, Thomas Arundel. Henry Howard, who was carrying crippling debts run up by his father before the Commonwealth period, was all too happy at the Royal Society's effectively taking over upkeep of Arundel House. In January 1667, encouraged apparently by his close friend John Evelyn, he presented the Library to the Royal Society as a magnificent New Year gift, thus absolving himself of responsibility for its upkeep also. The Royal Society minutes for 2 January record:

Mr Henry Howard of Norfolk presented the society with
the library of Arundel-house, to be disposed thereof by
them as their property, desiring only, that in case the society
should come to fail, it might return to Arundel-house; and
that this inscription 'ex dono Henrici Howard Norfolciensis'
might be put upon every book given them; he allowing also
the liberty of changing those books, that were double, or
such as were not for the society's purpose, for others; which
exchanged books were to be marked likewise with the same
inscription.

From the outset the arrangements for running the Library appear
to have been haphazard. A number of Royal Society members, and
notably Hooke himself (who, as Curator, held the Library key),
took advantage of their easy access to the Arundel Library to make
extensive use of it, extending to long-term borrowing of a number
of books (there was undoubtedly book-stealing too, on a grand
scale). By 1672 grave concern was being expressed that no one was
taking proper care of the Library.

Hooke's own library was an impressive collection of up-to-the-
minute specialist publications in a wide range of fields, such as we
might expect for a working scientist with varied interests. He bought
enthusiastically from booksellers, and attended the many London
book auctions regularly. On occasion he would spend serious money
on a rare book, or a contemporary book difficult to obtain on the
open market (particularly books on architecture). He exchanged
books with friends, and bought privately via the coffee-houses of
which he was a regular frequenter. The inventory published for the
1703 auction of his books after his death ran to several thousands
of books. One might think that he had no need to borrow other
people's books, but we all know that bibliophiles are often inveter-
ate 'borrowers'. The charge that Hooke was borrowing without
authorisation from the Arundel Library is based on his own diary
entries. On 9 April 1673, for instance, he brought home 'from

Arundell library', '3 volumes of Vasari of the Lives of the Painters, Cesari Ripa of Iconologia, Rubens Life etc.'.

Things seem to have come to a head in early 1677. Howard expressed his concern at the state of the Library, which was still at Arundel House, as was the Royal Society. At the 25 January Council meeting, 'Mr Oldenburg acquainted the council from the earl marshal, that his lordship was desirous, that the library at Arundel House given by him to the Royal Society might be better looked after.' Howard had also apparently decided to withdraw from the Library gift all books on heraldry and associated subjects, in order to give these instead to the College of Arms (the association of heralds, of which he was the titular head). The Royal Society minute continued:

> as also, that Henry Howard, Earl Marshall, should be glad to
> have those books of that library delivered to him, which he had
> reserved out of it to himself at the time of the donation
> thereof, viz. books of heraldry and genealogy. It was ordered
> hereupon, that Sir John Hoskyns, Mr. Evelyn, Mr. Hill, and
> Mr. Oldenburg, or any two or more of them, should be a
> committee to attend the earl marshal, to deliver to his lordship
> such books as he had reserved to himself out of the Arundelian
> library; as also to secure that library from damage.[50]

John Evelyn recorded in his diary:

> I was called again to London to wait again on the Duke of
> Norfolk who having at my request only, bestowed the Arundel
> Library on the Royal Society, sent me to take charge of the
> books and remove them: only that I would suffer the Heralds'
> Chief Officer Sir William Dugdale to have such of them as
> concerned heraldry and the marshall's office, as books of
> Armourie and Geneologies (the Duke being Earl Marshall of
> England). I procured for our Society besides printed books,
> neer 700 manuscripts, some in Greek of great concernment;

the printed books being of the oldest impressions, are not the less valuable, I esteem them almost equal with the manuscripts.[51]

Hooke's diary, meanwhile, records the drama that was going on behind the scenes. On 25 January, the President of the Royal Society summoned Hooke, and accused him of neglecting the Arundel Library and misusing his access to it: 'Brounker would have had the key of Arundel library taken from me.'[52] The following day Hooke was called again before the President, who instructed Hooke to get the Library cleaned at once, and to put back his own long-term 'loans':

> To Brounker. He orderd me to returne home books to Arundell library. With Ned, Harry and woman thither. Made fire and cleansed the books and Roome. Sir J. Hoskins there I returnd to Arundell Library, Palace de Genoa 2 vol., Fabius Columna 2 vol. 4d., Hortus Farnesianus I, Acosta, Linscoten, Palladio, Vasari 4 volumes, Le matre, Cesari Ripa Iconologia, Vita di Titiano Aretinus, Morus, Ureses perspective, and some mapps.[53]

So the books went back, including, perhaps, the three volumes of Palladio (although the 1703 sale catalogue of Hooke's books contained such a three-volume Palladio set), and apparently Hooke succeeded in convincing Brouncker and Howard that he was still a suitable librarian. On 3 March he 'spoke with Lord Marshall about Library and Sir J. Laurence his business of the Key'.[54] Subsequent entries in the minutes of the Royal Society, show that Hooke was authorising six-month loans of manuscripts to specified members, and that he was responsible for recording the loan and ensuring the eventual return of the item.

In August 1678, the Royal Society finally vacated Arundel House, and Howard 'renewed the declaration of his gift formerly made to the Society of the Arundelian library; and also gave his

10. Palladio temple from his influential treatise on architecture

consent and direction for the removal thereof into the possession of the Society'. (The Society promptly sold the manuscripts to the Bodleian Library in Oxford.) It still remained, however, to separate out the heraldry books. This task was placed in the hands of the elderly Sir William Dugdale, who was particularly well qualified to do the job. He was a senior official of the College of Arms himself, and a considerable book collector.

On the other hand, it may have not been particularly sensitive of Howard to assign Dugdale to the task. He had been one of those who had lost his entire personal library in the Great Fire ('Sir William Dugdale hath lost about 1000 pounds in books,' Pepys had noted). Perhaps Dugdale found himself including among the books that would not go to the Royal Society some that particularly appealed to himself, or that he had owned before his library burned. The lengthy list he drew up included many volumes that could not be construed as relevant to heraldry at all. Hooke noted this, and there followed a quarrel in a coffee-house between himself and Sir Thomas St George, Dugdale's assistant, whom he accused of dishonestly acquiring Arundel Library books for himself. The suggestion was that an inflated list of books was being submitted to the Royal Society, from which a significant number of volumes would be quietly extracted and retained before the *bona fide* heraldry books were passed to the College of Arms.[55] Finally Howard, whose suspicions had been aroused, had the lists checked at both ends, to ensure that they tallied.

Secretary to the Royal Society

The first secretary of the Royal Society, Henry Oldenburg, came to that job by accident. Born in Bremen, in what today is Germany, he came to England in 1653 as envoy for the Free City of Bremen, though to judge by his excellent English he had probably spent

I. Specimen of cocoa plant from Sloane's collection

2. Sheet of Maryland plants from Sloane's collection

3. Teniers's vivid image of an alchemist (or early chemist)

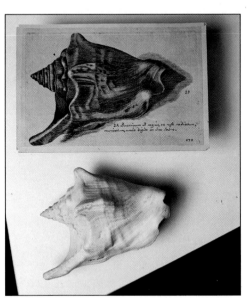

4 & 5. Engravings, with the shells observed, produced by Lister's daughters for his *Historia Conchyliorum*. The shells form part of Sloane's collection

6. Contemporary Dutch painting of the Great Fire of London

OPPOSITE
7. Reconstruction of Newton's reflecting telescope from
the original design

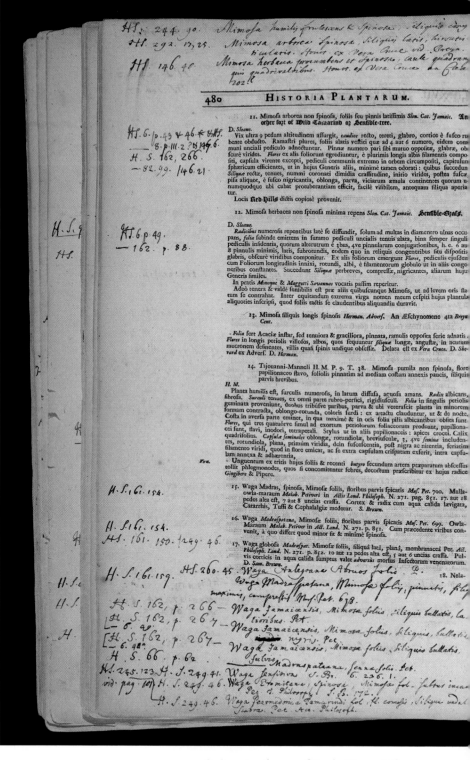

[handwritten marginal annotations]
HS: 244. 90. Mimosa humilis, frutescens & spinosa, siliquis ...
Hl. 292. 17, 25. Mimosa arborea spinosa, siliquis latis, hirsuti...bularis. ... ex Vera Cruce vid. Breyn.
HS. 146. 40. Mimosa herbacea procumbens et spinosa, caule quadran... quadrivalvibus. ... ex Vera Cruce ... 102 li

11. Mimosa arborea non spinosa, foliis seu pinnis latissimis *Sloan. Cat. Jamaic.* An other sort of Wild Tamarind or Sensible-tree.

D. Sloane.
Vix ultra 9 pedum altitudinem assurgit, *caudice* recto, tereti, glabro, cortice è fusco rubente obducto. *Ramastri* plures, foliis alatis vestiti quæ ad 4 aut 6 numero, eidem communi unciali pediculo adnectuntur. *Pinnæ* numero pari sibi mutuo oppositæ, glabræ, obscurè virides. *Flores* ex aliis foliorum egrediuntur, è plurimis longis albis filamentis compositi, capsula virente excepti, pediculi communis extremo in orbem circumpositi, capitulum sphæricum efficientes, ut in hujus Generis aliis, minimè tamen odorati: quibus succedunt *Siliquæ* rectæ, tenues, nummi coronati dimidia crassitudine, initio virides, postea fuscæ, pisa aliquot, è fusco nigricantia, oblonga, parva, viciarum æmula continentes quorum unumquodque ubi cubat protuberantiam efficit, facilè visibilem, antequam siliqua aperiatur.
Locis Red-Hills dictis copiosè provenit.

12. Mimosa herbacea non spinosa minima repens *Sloan. Cat. Jamaic.* Sensible-Grass.

D. Sloane.
Radicibus numerosis repentibus latè se diffundit, solum ad multas in diamentro ulnas occupans, *folia* subinde emittens in summo pediculi unciali tenuis alata, bina semper singuli pediculis insidentia, quorum alterutrum è 3bus, 4ve pinnularum conjugationibus, h. e. 6 aut 8 pinnulis minimis, laris, subrotundis, eodem quo in reliquis congeneribus situ dispositis glabris, obscurè viridibus componitur. Ex aliis foliorum emergunt *Flores*, pediculis ejusdem cum Foliorum longitudinis innixi, rotundi, albi, è filamentorum globulo ut in aliis congeneribus constantes. Succedunt *Siliquæ* perbreves, compressæ, nigricantes, aliarum hujus Generis similes.
In pratis *Minque* & *Maggotti Savannas* vocatis passim reperitur.
Adeò tenera & valdè sensibilis est præ aliis quibuscunque Mimosis, ut ad levem oris statum se contrahat. Inter equitandum extrema virga nomen meum cespiti hujus plantulæ aliquoties inscripsi, quod foliis ractis se claudentibus aliquandiu duravit.

13. Mimosa siliquis longis spinosis *Herman. Adverf.* An Æschynomene 4ta Breyn Cent.

. *Folia* ferè Acaciæ instar, sed tenuiora & graciliora, pinnata, ramulis opposita serie adnatis. *Flores* in longis petiolis villosos, albos, quos sequuntur *siliquæ* longæ, angustæ, in acutum mucronem desinentes, villis quasi spinis undique obsessæ. Delata est ex *Vera Cruce.* D. Sherard ex Adverf. D. Herman.

14. Tsjouanni-Manneli H. M. P. 9. T. 38. Mimosa pumila non spinosa, flore papilionaceo flavo, foliolis pinnatim ad mediam costam annexis paucis, siliquis parvis brevibus.

H. M.
Planta humilis est, surculis numerosis, in latum diffusa, acuosa amans. *Radix* albicans, fibrosa. *Surculi* tenues, ex omni parte rubro-persici, rigidiusculi. *Folia* in singulis petiolis geminata proveniunt, duobus tribusve paribus, parva & ubi veterascit planta in minorem formam contractæ, oblongo-rotunda, coloris surdi: ex atractu clauduntur, ut & de nocte. Costa in aversa parte eminet, in qua maximè & in oris folia pilis albicantibus obsita sunt. *Flores*, qui tres quatuorve simul ad exortum periolorum foliaccorum prodeunt, papilionacei sunt, flavi, inodori, tetrapetali. Stylus ut in aliis papilionaceis: apices crocei. Calix quadrifolius. *Capsulæ seminales* oblongæ, rotundiolæ, breviusculæ, 3, 4ve *semina* includentes, rotundiola, plana, primùm viridia, dein fuscescentia, post nigra ac nitentia, seriatim filamento viridi, quod in flore emicat, ac se extra capsulam crispatam exserit, intra capsulam annexa & adhærentia.
. *Unguentum* ex tritis hujus foliis & recenti *butyro* secundum artem præparatum abscessus tollit phlegmonodes, quos si concomitentur febres, decoctum præscribitur ex hujus radice Gingibere & Pipere.

15. Waga *Madras*, spinosa, Mimosæ foliis, floribus parvis spicatis *Muf. Pet.* 700. Mullaowla-maraum *Malab.* Petiveri in *Actis Lond. Philosoph.* N. 271. pag. 851. 17. aut 18 pedes alta est, 7 aut 8 uncias crassa. Cortex & radix cum aqua calida lævigata, Catarrhis, Tussi & Cephalalgia medetur. S. Brown.

16. Waga *Madraspatana*, Mimosæ foliis, floribus parvis spicatis *Muf. Pet.* 699. Owla-Maraum *Malab.* Petiver in *Act. Lond.* N. 271. p. 851. Cum præcedente viribus convenit, à quo differt quod minor sit & minimè spinosa.

17. Waga globosa *Madraspat.* Mimosæ foliis, siliquá latá, planá, membranaceá Pet. *Act. Philosoph. Lond.* N. 271. p. 852. 10 aut 12 pedes alta est, 5 aut 6 uncias crassa. Hujus corticis in aqua calida sumptus valet adversùs morsus Insectorum venenatorum. D. Sam. Brown.

18. Nela-

[handwritten left-margin references, selected]
H.S. 6. p. 43 & 46
8. p. III. 2 ... 146. 6.
H. S. 162, 266.
—82. 99. 146. 21
H. S. 6 p. 49
—162. p. 88
H. S. 6. 154.
H. S. 6. 154.
H.S. 161. 150. 249. 46
H. S. 161. 159.   HS. 260. 45 Waga ...
H. S. 162, p. 266 —
H. S. 162, p. 267 —
H. S. 66. p. 62
H.S. 245. 123. H. S. 249. 41. Waga sensitiva ...
H. S. 249. 46. Waga ...
H. S. 249. 46. Waga ...

8. Annotated copy of Ray's *History of Plants*

18. Nela-tali H. M. P. 9. T. 18. Æfchynomene flore papilionaceo, Ferri equini filiquis.

H. M.

Nafcitur in udis & paludofis, & paffim inter *Oryzam*. *Radix* fibrofa, denfè capillata, carne fungofa, denfa, molli, levi, candida, corculo pennæ haud diffimili, ac in teneris fibris corticula flava & pilis flavefcentibus obducta. *Caules* teretes, glabri, virides, intus carne in totum fungofa, albicante; teneriores ex una parte rubefcunt. *Folia* è plurimis pinnarum parvarum, oblongo-rotundarum, exteriùs in medio venâ ftriatarum, viroris fufci, denfè ftipatarum conjugationibus componuntur, nullo in extremo impari foliolo. Si folia hæc paulò preffiùs ftringantur, ftatim fe fcutiformiter ad fe invicem in duabus adverfis feriebus claudunt, ut & in furculis cum radice extractis feu avulfis. *Flores* ad exortum petiolorum florum in pedunculis videntur tenuibus, papilionacei, parvi, flavifculi, tetrapetali, exterius foliolum maximè interius rubefcit, circa oras flavefcens, duo interiora flaviufcula funt, quartum viridi-albicans. *Siliquæ* oblongæ funt, planæ, paululum inflexæ, cortice primùm viridi, poft ruffo-fufco & fcabro. *Semina* quæ per corticem extuberant, parva funt, oblonga, faporis leguminofi, in longitudine filiquarum in propriis loculis difpofita, futuris intermediis in cortice exuberantibus diftincta : Color fpadiceo-fufcus eft ac nitens.

Lavamentum ex hac planta conficitur antivulnerarium.

19. Todda-Vaddi H. M. P. 9. T. 19. Herba viva Acoftæ cap. 5. Totta-Vari *Zanon.* feu Herba Mimofa *Malabarenfium*, qui tamen fuam *Totta-vari* ab Herbâ viva *Acoftæ* differre exiftimat.

Hujus Defcriptionem accuratam videfis hoc in loco. Nos alibi de Herba viva *Acoftæ*.

20. Herba viva latifolia *Javanica Breyn.* Prod. 2. Herba fentiens vivens, five Æfchynomene foliis & floribus longis pediculis, unico thyrfo infidentibus *Amboinica Rumphii* in *Mifcell. curiof.*

E *Java* infula ad D. *Breynium* tranfmiffa fuit per Clariff. *Cleyerum.*

21. Mimofa hirfuta pumila, flore purpurafcente, quæ tacta ramulis non modo clauditur, fed humi profternitur. Mimofa humilis herbacea *Ind. Orient.* femine vulgaris æmulo : Chaddai-lackaree *Malabarorum Pluk. Mantiff.*   *H.I. 9 2. 94.*

22. Mimofa humilis quadripennis, glabra, procumbens guttis fanguineis maculata, ex infula *Ceylon. Ejufd. ibid.* In Horto Epifcopali Fulhamiæ prope *Londinum* colitur.   *H.S. 88. 75. / H.S. 9 2. 94.*

23. Mimofa *Indica*, filiquis plurimis fimul junctis, latis, brevibus, flore parvo globofo : Varainchuddee *Malabarorum Pluk. Mantiff.*

24. Mimofa pumila *Ceylonenfis*, erecta, multifolia, pinnulis fingulis maculâ purpureâ utrinque notatis *Ejufd. ibid.*

25. Mimofa non fpinofa, Coluteæ foliis, floribus ftamineis, filiquâ latâ, fufcâ, brevi, & lignofâ : Erewetamaram *Malabarorum Ejufd. ibid.*   *H.S. 93 - 5. ? /94 - 18. - 92. 94.*

26. Niri-Todda-Vaddi H. M. P. 9. T. 20. *Herba Cofta Zeylanica*, filiquis latis compreffis minoribus *Breyn.* Cent. 1. fol. 47. ex fententia D. *Commelin.*

Nafcitur in aquis dulcibus, aquæ cum caulibus innatans, femper florens, præfertim tempore pluviofo. *Radices* ex caulibus ordinatim non indè prodeunt, maxima parte prope prope inflexuras, fûntque condenfæ ac breves fafciæ, ex tenuibus fibris condenfatis ac fibi mutuo implexis conftante, coloris Perfici. *Caules* aquis innatantes rotundi, duriufculi, viridi-clari, aft Soli expofiti purpurafcentes, fungofi intus & albicantes, fiftulofi, ad corticem lignofiores, angulis in contrarias partes alternatim inflexi, proximè circa inflexuras fungofa, fpongiofa, humida, albicante tunica circumveftiti. *Folia* bina ac bina in periolis mediis, ac ad exortum exuberantibus ac rubefcentibus, proveniunt, forma angufto-oblonga, non tam lata quàm illa *Todda-Vaddi*, in fummitate rotundiola, latitudine uniformia, mollia, tenuia, plana & valde lenia, coloris hyalini, fubtus venulam habent fubtilem ad fui exortum nonnihil eminentem, folium in duo inæqualia dividens latera. *Folia* in hac non in eodem plano ut illa primæ fpeciei, fed diverfis planis fibi mutuo parallelis, petiolum medium oblique fecantibus, & verfus fummitatem nonnihil inflexis jacent, tali modo fita ut aperta cum exteriori fuperficie aquæ obverfa fint ; ex attactu clauduntur, ut & vefperi ac internoctu. *Flores* fpicatim & condenfi in orbem proveniunt in fummitate propriorum petiolorum, qui rigidi, rotundi, nonnihil inflexi, coloris Perfici diluti, & ex inferioris caulis vel folitarii, vel cum caulibus prodeunt ; fûntque inferiorès plurium foliorum valde oblongo-anguftorum & crifpatorum, coloris in totum flavi, in calice nullis ftamina vel filiquofum germen edentes, fuperiores pentapetali, virides, cufpidati. Hi 10 habent *ftamina* longiufcula, albicantia, cum fuis ftylis,ftriatis, & obliquè furrectis *apicibus*, inter quæ *ftylus* eminet longus ex filiquofo viridi germine in umbilico floris oriens. *Siliquæ* parvæ funt, ex umbilico floris ortæ, planæ, compreffæ.

used as partial index to Sloane's herbarium

9. Secretary to the Royal Society, Henry Oldenburg, with his watch

time in England in the 1640s. His mission was to negotiate the release of the cargo of a ship owned by a citizen of Bremen, which had been captured by English men-of-war (laden with 'four tons of Nantes wine and twelve quartols of brandy wine'). While he was there, however, Bremen was forcibly absorbed into the territories of the King of Sweden, and Oldenburg's diplomatic post lapsed. Instead he became tutor to the son of Boyle's sister, Lady Ranelagh, with whom he was travelling on the Continent when Charles II was restored to the throne. On his return he moved to Boyle's employment, performing the services of editor, translator and informant on all that was going on, intellectually and politically, Europe-wide.[56]

Oldenburg was never a scientist. Boyle brought him into science, introduced him into the Royal Society, and obtained for him the job

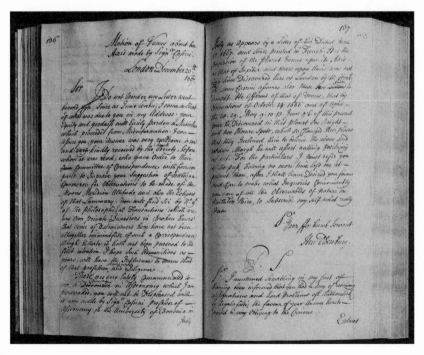

11. Transcription of Oldenburg correspondence in the Royal Society letter-book

12. Secretary to the Royal Society, Henry Oldenburg, with his watch

of secretary, which he shared initially with John Wilkins. From the 1660s, Oldenburg was a professional information-gather, or 'intelligencer', first on behalf of Boyle and then, increasingly, on behalf of the Royal Society and the London scientific community in general. Martin Lister described how he worked:

He held Correspondence with seventy odd Persons in all Parts
of the World, and those [to] be sure with others. I ask'd him
what Method he used to answer so great a variety of Subjects,
and such a quantity of Letters as he must receive weekly; for I
knew he never failed, because I had the Honour of his
Correspondence for Ten or Twelve Years. He told me he made
one Letter answer another, and that to be always fresh, he never
read a Letter before he had Pen, Ink and Paper ready to answer
forthwith; so that the multitude of his Letters cloy'd him not,
or ever lay upon his Hands.[57]

In the early years, Oldenburg's extraordinary correspondence
dominated the early Royal Society's activities, shaping its concerns
and interests. It was Oldenburg who decided which letters would be
brought to the attention first of the Council and then of members

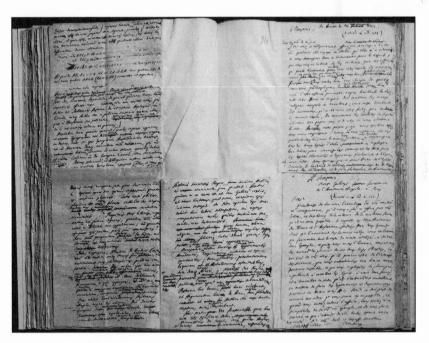

13. Original Oldenburg letters in the Royal Society's archives

at their regular meetings. It was Oldenburg who recorded proceedings at the meetings, and then published them in the Society's *Transactions*.

At the end of the second Dutch war, Oldenburg was arrested on suspicion of having been a spy during the hostilities. Evelyn records in his diary:

> August 8, 1667: Home, by the way visiting Mr Oldenburg now
> close prisoner in the Tower, for having been suspected to write
> Intelligence &c: I had an order from my Lord Arlington,
> Secretary of State, which made me to be admitted; this
> Gentleman was Secretary to our Society, & will prove an
> innocent person I am confident.[58]

The charge was not so surprising. Oldenburg's last trip out of England before he settled there, in 1661, had been to the Dutch Republic, where he had met Huygens for the first time. He had continued corresponding with Dutch and French scientists throughout the war, when both the Dutch Republic and France were hostile powers. In particular, he kept up a lengthy, detailed correspondence with Huygens – a Dutch national, resident in France – extending to descriptions of military matters like troop movement and engagement.

No wonder Oldenburg was so sensitive to the charge of spying levelled at him by Hooke at the height of the dispute between Hooke and Huygens over priority for the invention of the balance-spring watch in 1675–6.[59] On 11 October 1675 Oldenburg wrote to Huygens in some agitation:

> Mr Hooke, having known that you had given me permission
> to make use of the benefit which you could claim of a patent
> for your watch in this country, has been so rash and
> shameless as to say publicly that you gave me this permission
> as a reward for having divulged to you his invention, adding
> with the greatest effrontery in the world that I am your spy

here for communicating everything of note found out here, and that I wanted to defraud him of the profit of his invention.[60]

It must surely give us food for thought, as we survey the paper record of early science in England, to consider that its systematic and solid foundation was laid by a German Protestant, with Low Countries sympathies, whose pan-European correspondence more than once earned him the reputation of a spy.

EPILOGUE

IN 1968 A scandal erupted at Harvard University. The story got out that Harvard University Press was refusing to publish *The Double Helix*, the thrilling inside story of the scientific discovery of DNA (deoxyribonucleic acid). The book was written by James Watson, one of the team who had made this fundamental breakthrough in molecular biology. The decision not to publish was sufficiently remarkable to make the national press. In the *New York Times*, a correspondent explained why the unusual decision had been taken: 'The university halted plans for publication when Drs F. H. C. Crick and M. H. F. Wilkins, the two men who shared the Nobel Prize with Dr. Watson for this work, voiced protests.'[1]

Francis Crick and Maurice Wilkins had apparently objected to parts of Watson's manuscript, which they were shown in draft, not because Watson's account was historically inaccurate, but on the grounds that it was gratuitously hurtful in its characterisations of, or offhand remarks about, people involved in the story. According to the *New York Times* again: 'The book tells the story in highly personal terms, describing the idiosyncrasies of the principals, their quarrels and friendships.'

Watson removed, or watered down, a number of offending passages.

One other person closely involved with the story had not been able to comment on the manuscript. Maurice Wilkins's co-worker Rosalind Franklin, who had probably come closest to solving the problem of the structure of DNA before Crick and Watson, had died of cancer in 1958, at the age of thirty-seven. Her early death had already prevented her sharing in the Nobel Prize awarded to her fellow scientists in 1962, since that prize is not bestowed posthumously. Now she could not respond to Watson's gratuitously offensive remarks about her personality and her work. Watson acknowledged the awkwardness of this, and added an 'epilogue' to *The Double Helix* in which he conceded that he had been wrong in some of his less flattering early impressions of Franklin, as recorded in his book.

The furore surrounding the publication of *The Double Helix* did not, however, in the end hinge on hurt feelings or wounded pride. At issue was the account Watson gave of the way scientific discoveries are made. Where conventional twentieth-century accounts made the scientist an omniscient sage, Watson made him more like a carpet-bagging adventurer, driven on by his competitive spirit and the desire to win at all costs.

Fellow scientist André Lwoff, reviewing *The Double Helix* for *Scientific American* in July 1968 (the book was eventually published by another press) had no doubt that Watson had 'told it how it was', so far as scientific practice was concerned: 'For the first time all the steps and circumstances of a major contribution to science are described with precision and accuracy.'[2]

Robert L. Sinsheimer, reviewing it in *Science and Engineering*, thought less well of the author and the more personal parts of the book than Lwoff, but he too endorsed the account of the way the discovery was made:

In reality this is two books. One is an account — lucid, honest, suspenseful — of the scientific events that led to the deduction

of the molecular structure of DNA, which at one stroke
provided a clear chemical basis for the results of 50 years of
genetics and at the same time constituted the central support
about which the whole structure of molecular biology could be
built. Because it is in many ways a typical story of scientific
discovery – with false trails, the fortuitous combinations of
ideas, the ex-post-facto-obvious nature of the solution – with
all the drama heightened by the importance of the goal – it
could well serve as a model text for initiation of the young,
were it not for the second book.[3]

The 'second book', according to Sinsheimer, was lurid, malicious
and mean-spirited, a chronicle of two ambitious and arrogant
young men on the make. Having chastised the author for presenting the scientist and his pursuit in such terms, however, Sinsheimer
concluded:

It is perhaps an interesting psychological question, if indeed
these two books – the components of *The Double Helix* – are not
in themselves complementary; if indeed the structure of DNA
would have been discovered in this way had it not been for both
the slanting brilliance and the skewed personality of J. D.
Watson. Probably not.[4]

As a narrative available for scrutiny by the historian of science,
Watson's *The Double Helix* is the modern equivalent of Hooke's diary.
It is a highly personal narrative account of the everyday lives of
research scientists, and the complicated ways in which they move
from humdrum laboratory work towards the solution of a problem
identified as of major importance to the community at large. What
must strike us, at the end of *Ingenious Pursuits*, is how similar the tales
are that Hooke and Watson tell: tales of casual encounters in coffee-
houses, overheard remarks and data encountered fortuitously, results
read about in the proceedings of adjacent but notionally unrelated
fields. There are brash remarks, rash promises of success, mistakes

hastily withdrawn and reworked, personality clashes that hinder breakthroughs and glittering prizes for a few at the end of the day. Some of the participants are disinterested pursuers of knowledge, while others are ambitious and unscrupulous, determined to be the first to solve the modern equivalent of the longitude problem.

After the Second World War, a number of scientists in Europe and America who had previously worked in physics turned their attention instead to biological material, and began working on the molecular structure of the genes that transmit hereditary characteristics in living things. Their inspiration was a book entitled *What is Life?*, by the great physicist Erwin Schrödinger (first published in 1944), in which he argued that the underlying structures in biology were susceptible to the same kinds of molecular analysis as the rest of the physical world.[5]

This new field of scientific investigation became known as 'molecular biology', and in the years immediately post-war, knowledge in this area developed at the same kind of breakneck speed as did astronomy and mechanics in the second half of the seventeenth century. By the early 1950s these molecular biologists had focused their attention on DNA as the likely carrier of human genetic material. The English scientist W. H. Bragg and his son Lawrence pioneered the field of X-ray crystallography, built on techniques of photography at the minute molecular level, which combined theory and advances in technology and techniques for using it that recall late seventeenth-century microscopy.

The leading practitioners in this field around 1950 were Maurice Wilkins and his co-worker Rosalind Franklin, at King's College, London. James Watson first saw one of Maurice Wilkins's X-ray diffraction photographs of the DNA molecule at a conference in Naples in the spring of 1951. He had not yet met Francis Crick, and was working as a young post-graduate in a Copenhagen biochemistry laboratory, on material and with techniques that did not greatly interest him. When he saw Wilkins's photographs, however,

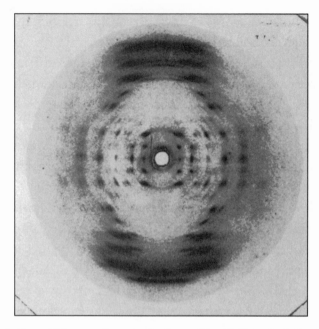

I. X-ray diffraction photograph of crystalline DNA in the A form

he immediately recognised that if genes crystallised, as the photographs suggested, there was a possibility of discovering their fundamental structure by crystallography. He applied to his American funding organisation to switch his research to the Cavendish Laboratory in Cambridge, England, where Max Perutz was known to be working on the minute structure of large biological molecules. His application was successful, and he moved to England in the autumn of 1951.

'From my first day,' writes Watson, 'I knew I would not leave Cambridge for a long time. Departing would be idiocy, for I had immediately discovered the fun of talking to Francis Crick.'[6] Crick was more than ten years older than Watson, but because the Second World War had interrupted his graduate research he was still without his doctorate. He was as voluble and excitable as Watson was awkward and shy. As Watson puts it, with an admiration laced with his customary irony: 'Francis' brain was a genuine asset. Though a few dissidents still thought he was a laughing talking-machine, he nonetheless saw problems through to the finish line.'[7]

Above all, Crick was a connector of information gathered and processed on a grand scale, a detector, at speed, of the complicated interrelatedness of one parcel of knowledge and another. He began his day by scouring the latest scientific journals, not simply in his own field, but in as many related fields as possible. He was full of ideas, though easily bored. He loved scientific debate, and would engage anyone in the Cavendish in heated discussion of the topic of the moment until they, but not he, were exhausted. Watson, who was as conversationally indefatigable as Crick once his interest was fired up, became Crick's regular coffee-break and lunchtime companion in the lab, and the two began to meet increasingly often outside the lab, where they continued their scientific discussions whenever they were not distracted by their other shared obsession: the opposite sex.

What Watson calls 'a long time' was in fact a period of exactly two years: by the end of 1953 Watson had moved back to the United States, to work at Cal Tech. What is really remarkable is the speed at which Crick and Watson homed in on the heart of the problem of DNA, then solved it. It was a propitious time to crack this particular problem, just as the moment was finally right to solve the longitude problem after 1700. The key technology and vital techniques were available. In 1951, at Cal Tech in the United States, Linus Pauling had discovered that the basic structure of the protein molecule – a long chain molecule closely related to DNA – was helix-shaped. His original (and, in crystallography terms, thoroughly unconventional) technique for arriving at this conclusion involved a combination of guesswork and physical model-building. It was this same technique that Crick and Watson would use to solve the structure of DNA.[8]

It was Pauling's discovery of the helical structure of large protein molecules that took up a lot of Crick and Watson's lab discussions, together with a helter-skelter exchange of expertise from their respective specialist fields: Crick quizzed Watson on biochemistry; Watson quizzed Crick on crystallography. Watson reports that he soon understood from Crick 'that Pauling's accomplishment was the product of common sense, not the result of complicated mathematical reasoning':

> The key to Linus' success was his reliance on the simple laws of structural chemistry. The alpha-helix had not been found by only staring at X-ray pictures; the essential trick, instead, was to ask which atoms like to sit next to each other. In place of pencil and paper, the main working tools were a set of molecular models superficially resembling the toys of preschool children. We could see no reason why we should not solve DNA in the same way.[9]

They set about trying. Crick and Watson played around with home-made models, they consulted vigorously with Wilkins and Franklin in London about the X-ray diffraction photographs, while carefully concealing the reason for their interest, and Watson began taking X-ray diffraction photographs himself at the Cavendish, taking advantage of the state-of-the art, powerful rotating anode X-ray tube just assembled there. Then, in December 1952, Linus Pauling's son Peter, with whom they shared an office at the Cavendish, announced that his father had written to tell him he had solved the structure of DNA.

Pauling turned out to have made a mistake; his model was wrong. But the letter galvanised Crick and Watson into action. With a single-minded determination entirely reminiscent of Hooke's pursuit of the balance-spring secret following Huygens's announcement that he had solved the problem of a longitude watch, they redoubled their own efforts to solve the problem themselves.

A quarrel with Franklin gave Crick and Watson the final, vital piece of data they needed. Visiting her lab in London early in January 1953, Watson, with his customary lack of tact, lectured her

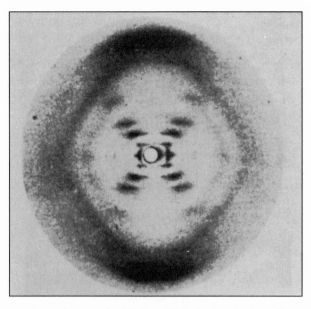

2. X-ray photograph of B form DNA by Rosalind Franklin

on the need for her to learn some theory to interpret her outstanding X-ray diffraction photos of DNA. She exploded with
understandable anger, and he fled. In the safety of her co-worker
Wilkins's laboratory the two men moaned about the perils of working with women, and Wilkins produced Rosalind Franklin's latest,
astonishing photograph. Watson saw immediately that it gave him
the information he needed:

> The instant I saw the picture my mouth fell open and my pulse
> began to race. The pattern was unbelievably simpler than those
> obtained previously. Moreover, the black cross of reflections
> which dominated the picture could arise only from a helical
> structure.[10]

As it happens, Wilkins also told Watson what Franklin believed
to be the structure of the DNA molecule on the basis of her photograph – in which she turned out ultimately to be entirely correct.
Watson, with his bruised ego, was hardly likely to agree at this

moment: 'I remained skeptical, for her evidence was still out of the reach of Francis [Crick] and me.'[11]

Watson rushed back to Cambridge with a precious sketch – on the edge of his newspaper – of Franklin's crucial photograph. He and Crick built models and argued about them with increasing vigour. They formulated convincing theories in the evening, and demolished them again the next morning. In two months they had it: in March 1953 they produced their double helix for scrutiny by the scientific community. Its simplicity was entirely convincing. A letter to *Nature* followed. The paper explicating the discovery ('The

3. Watson and Crick in front of their DNA model

structure of DNA', authored jointly by J. D. Watson and F. H. C. Crick) was published in the *Cold Spring Harbor Symposia on Quantitative Biology* that same year. In 1962 Crick, Watson and Wilkins shared the Nobel Prize 'in Medicine or Physiology'.

I have recapitulated the ingenious pursuit of DNA in a certain amount of detail because in Watson's version of events, largely borne out by Crick's few, much more discreet comments, it so closely resembles any number of the stories in this book. When the tales are told in this way, we put back the people into the laboratory, and the laboratory into its wider community. That produces a kind of anthropology of science, a study not just of the collections of data, the laboratory reports and the third-person accounts of important scientific proofs and discoveries, but of the interlocking of minds and behaviour, in a social setting, out of which fundamental new developments come and are reapplied for the benefit of the community from which they came.[12]

We have seen that the pursuit of science in the seventeenth century was an engaged, imaginative and even adventurous affair. It brought together creative talents of all kinds, from all walks of life. The collisions of intellects sometimes produced unexpected results; sometimes they simply concentrated their combined attention on a problem crying out to be solved. Between 1650 and 1750 science galvanised an entire continent, driving knowledge forward at an astonishing speed. It broke down international barriers, empowered the intellectually able but low-born, and broadened the horizons not just of the small circle of active practitioners but of entire communities. It heralded those everyday technologies that have allowed all of us, women and men alike, more and better opportunities to expand our individual understandings, increase our experience and, above all, learn.

The breakthroughs and new knowledge pivoted around new technologies. In the hands of skilled technicians, the telescope and the microscope, the precision watch and the micrometer gauge opened up new worlds to be explored, and ensured that the results of those explorations were meticulously measured and calibrated. Technology also tethered science in the real world. It grounded it in the 'Sincere Hand' and 'Faithfull Eye' of an observer. It prevented speculation from spiralling off into surmise. Franklin's X-ray diffraction photographs are the equivalent in this respect of Hooke's engraving of a microscopically enlarged thin slice of cork. Watson went on to become director of the Human Genome Project, the ultimate technical data-collection bank for human genetic material. Great science depends on remaining grounded in the real.

It is probably only since the 1930s that the scientist has come to figure in the public imagination as an almost entirely alien, morally irresponsible figure, whose interference with the natural world produced nuclear weapons capable of destroying the earth but who has failed to find a cure for cancer or Aids. Although Mary Shelley's *Frankenstein* came long before that, it was the Second World War that drove a wedge between ordinary citizens, with high hopes and values, and analysts and system-builders, who claimed their activities had no moral implications.

But scientists have never really been cut off from the rest of humanity in the way this suggests. As the story of DNA shows, scientists remain fully participants in 'ordinary life', with ordinary gifts and failings, ordinary passions and emotions. The problems that interest scientists are, inevitably, an extension of the problems that fascinate everyone, pursued systematically, with the help of all the technological and epistemological resources available. Like their precursors in seventeenth-century London, they gather some of their material during coffee-breaks, register a vital insight at a Royal Society meeting, or take a large step forward on the basis of an accidental encounter or a row in a laboratory.

As for the rest of us, we may not be able to follow the details of the scientists' proofs, but we are entitled to explanations we can understand. In return, scientists deserve to see us occasionally delighted by the beauty of the explanations they offer. We have been exploring the world around us scientifically for less than four centuries; we have only just started the process of understanding it. The questions have barely begun to be asked, but all of us will be implicated in the answers.

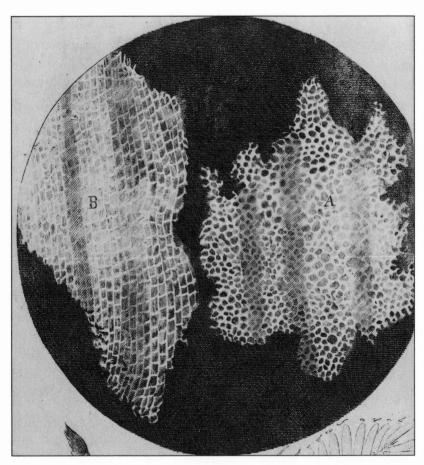

4. Hooke's engraving of a thin slice of cork, showing its cellular structure

CAST OF
CHARACTERS

Ashmole, Elias (1617–92)

Royalist antiquarian and Freemason. Rose to prominence after the Restoration of Charles II, redesigning ceremonies and hierarchies for the returning monarchy. Catalogued major collections of coins and medals at the Bodleian Library in Oxford, and for Charles II. Acquired John Tradescant's famous collection of curiosities and rare botanical specimens, which he had earlier catalogued, after the death of John Tradescant, and presented it to the University of Oxford, where it became the foundation for the Ashmolean Museum. Ashmole was the first recorded speculative Freemason, recording his admission to his Lodge in his diary.

Aubrey, John (1626–97)

Royalist gentleman and member of the Royal Society. Aubrey was a private scholar, antiquarian, and local historian who dabbled in astrology, and never worked for a living. He was a close friend of (amongst others) William Harvey, Thomas Hobbes, Robert Hooke and Christopher Wren; his acquaintances included Robert Boyle, Edmond Halley, Martin Lister, John Locke, Marcello

Malpighi, Isaac Newton, John Ray and Thomas Willis. Some of the biographical gossip he gathered was published in the 1680s in his *Brief Lives*.

Auzout, Adrien (1622–91)

French scientist, member of the Académie des Sciences in Paris. Foreign member of the Royal Society in London. Enthusiastic correspondent, whose letters to Oldenburg provide information on blood transfusions and other ground-breaking experiments.

Bacon, Francis (1561–1626)

James I's Lord Chancellor, disgraced and dismissed from office on charges of bribery. His *Great Instauratio* was a blueprint for the new sciences, and inspired several generations of practising natural philosophers. He was regularly invoked by the members of the Royal Society as the inspiration behind their activities.

Barrow, Isaac (1630–77)

Graduated from Trinity College, Cambridge, in 1644. Left England during the Commonwealth period, returning to a chair of Greek in Cambridge, and a chair of Geometry at Gresham College in London, in 1660. First Lucasian Professor of Mathematics at Cambridge from 1663 to 1669. Newton attended Barrow's lectures and addressed many important problems in physics as a result of Barrow's influence. Barrow resigned in 1669 so that Newton could take up the Lucasian Chair.

Bernoulli, Jacob (1654–1705)

Much-travelled Basel mathematician, eldest in a family of distinguished mathematicians. He corresponded with leading scientists of the day, including Hooke and Boyle. He returned to Switzerland and taught mechanics at the University in Basel from 1683, becoming Professor of Mathematics in Basel in 1687. In

1690 he challenged all comers to solve the problem of the curve of a freely hanging chain in the *Acta eruditorum*. He discovered the isochrone, the curve along which a body with uniform vertical velocity will fall.

Bernoulli, Johann (1667–1748)

Johann Bernoulli was the brother of Jacob Bernoulli. His elder brother tutored him in mathematics and Johann began to study Leibniz's papers. He was appointed to a chair in Groningen, the Netherlands, in 1695. After Jacob Bernoulli's death in 1705 he succeeded to his brother's chair in Basel. He remained in Basel for the rest of his life.

Blaeu, Joan (1595–1673)

Cartographer and map-printer in the Dutch Republic, employed by the Dutch East India Company for chart-making, and publisher of the beautiful, world-famous *Blaeu Atlas*. His publishing career ended when his printing premises were destroyed by fire.

Boucher, Charles (fl. 1675–90)

Friend of Edmond Halley at Cambridge, and fellow astronomical observer, whose journey to Jamaica inspired Halley to undertake his own trip to St Helena in 1676.

Boyle, Robert (1627–91)

Boyle was the seventh son of the first Earl of Cork, and never needed to work for a living. From 1656 he lived in Oxford where Hooke was a paid assistant in his laboratory. He made important contributions to physics and chemistry and is best known for Boyle's law (pressure × volume is a constant in an ideal gas), though Hooke did the experiments on which this was based. Boyle was a founding fellow of the Royal Society. His health was always poor.

Brahe, Tycho (1546–1601)

Of noble Danish extraction, Brahe was supposed to become a lawyer and diplomat, but instead became an astronomer. With financial help from the King of Denmark, Brahe set up a purpose-built observatory, on the island of Hveen in Copenhagen Sound. The observatory, called Uraniborg, was equipped with exceptionally large, accurate instruments, and with an alchemical laboratory in its basement. Twenty years later he quarrelled with the King and left Denmark. In 1599 he was appointed Imperial Mathematician to the Holy Roman Emperor, Rudolph II, in Prague. Kepler joined him as an assistant, and when Tycho died, Kepler succeeded him as Imperial Mathematician.

Breyne, Jacob (1637–97)

Botanist, son of a Dutch merchant who had settled in Gdansk. In 1678 Breyne published a flora of *Exotic Plants* (Gdansk, 1678).

Brouncker, William, Viscount (c.1620–84)

Viscount Brouncker of Castle Lyons, Brouncker was a founder and the first president (1662–1677) of the Royal Society of London. He graduated from Oxford in 1647, and was president of Gresham College, London from 1664 to 1667. Brouncker was a military man, with a particular interest in the science of firearms. His *Treatise on the Recoil of guns* was included in Sprat's *History of the Royal Society*.

Cassini, Gian Domenico (1625–1712)

Italian astronomer, Professor at the University of Bologna, lured to Paris by Colbert in 1669 to set up the Royal Observatoire and undertake a comprehensive programme of cartography, mapping first France and then the entire globe.

Cassini de Thury, Jacques (1677–1756)
Son of Gian Domenico Cassini, who also became Royal
Astronomer in Paris, and continued his father's work on a com-
plete map of France.

Cavendish, Margaret, Duchess of Newcastle (1624–74)
Prominent London aristocrat with a keen interest in science. She
wrote scientific treatises of her own, and enjoyed participating in
philosophical conversation. Her participation in the scientific
community in London was hampered, outside the salon, by her
sex.

Colbert, Jean-Baptiste (1619–83)
Louis XIV's minister of finance, Colbert was responsible for setting up the Académie Royale des Sciences in Paris, and ensuring
that it was adequately funded by the King. Also head of the military, he set the Académie to work on the complete mapping of
France. It was he who brought Cassini and Huygens to France.
From 1666 to 1683, Colbert spent an average of 87,700 livres a
year directly on the Académie – a vast sum. The other Académie
area of research strongly supported by Colbert was hydraulics.
The goal of this was to guarantee the supply and quality of water
for Versailles, and to design spectacular fountains for the royal
gardens.

Collins, John (1625–83)
First a clerk and then a seaman, Collins became a mathematics
teacher in London in 1649. From 1667 he worked as librarian
for the Royal Society in London, in addition to his other jobs.
Collins corresponded with Barrow, David Gregory, James
Gregory, Newton, Wallis, Borelli, Huygens, Leibniz, Tschirnhaus
and Sluze. He published books by Barrow and Wallis and left a
collection of 2000 books and large numbers of manuscripts.

Collinson, Peter (1694–1768)
Prominent London businessman who invested heavily in overseas expeditions both as an enthusiast for collecting, and because he was on the lookout for new commodities for the market.

Courten or Charleton, William (1642–1702)
Courten came of a prosperous business family, but his father had died insolvent. He studied natural sciences at Montpellier. On his return to London he became a considerable collector, establishing himself and his collection in a suite of rooms at the Temple. At his death he bequeathed his entire collection, valued at 50,000 pounds, to Sloane.

Descartes, René (1596–1650)
Born in France, Descartes was educated at the Jesuit college of La Flèche in Anjou. He received a law degree from the University of Poitiers. then enlisted in the military school at Breda. In 1618 he started studying mathematics and mechanics under the Dutch scientist Isaac Beeckman, and began to seek a unified science of nature. After two years in Holland he travelled through Europe. From 1620 to 1628 Descartes spent time in Bohemia (1620), Hungary (1621), Germany, Holland and France (1622–3). In Paris he made contact with Mersenne and began a lifelong correspondence with him. In 1628 Descartes chose to settle in Holland, where he spent the next twenty years. In 1649 Queen Christina of Sweden, with whom Descartes had corresponded, persuaded him to go to Stockholm, where he died of pneumonia.

Dixon, Jeremiah (1733–79)
Surveyor and astronomer, born in Bishop Auckland, County Durham In 1760 the Royal Society chose Charles Mason to go to Sumatra to observe the 1761 transit of Venus, and Mason

suggested Dixon should go as his assistant. An encounter with a French frigate delayed their final sailing so that they could not reach Sumatra in time. They therefore landed at the Cape of Good Hope, where the transit was successfully observed on 6 June 1761. In August 1763 Mason and Dixon signed an agreement with Thomas Penn and Frederick Calvert, Baron Baltimore, hereditary proprietors of the provinces of Pennsylvania and Maryland, to go to North America to help local surveyors define the disputed boundary between the two provinces.

Dugdale, William Sir (1605–86)
Antiquarian, author of the *Antiquities of Warwickshire*, and the definitive history of Old St Paul's (much admired by Wren and Hooke). He was a prominent member of the College of Arms, and became successively Chester Herald (1644), Norroy King of Arms (1660) and Garter King of Arms (1677). His daughter, Elizabeth, married Ashmole in 1668.

Evelyn, John (1620–1706)
Born in 1620 into a substantial Surrey landowning family whose fortunes were founded in gunpowder manufacture, John Evelyn travelled in Italy and France during the Commonwealth period. By the time he returned to England in 1652 he had become a committed intellectual, and thenceforth was an active participant in coffee-house life in London, and in the Royal Society (he never took a job). His gardens in Deptford were famous, and were substantially modelled on French and Italian gardens he had visited during his years abroad. His diary is one of the key sources for the period, though unlike Pepys he wrote it retrospectively, sometimes a considerable time after the events he recorded.

Faithorne, William (1616–91)
Engraver and publisher, who engraved the plates for many of

Boyle and Hooke's publications, and engraved the plates for Ray's *History of Fishes*. He replaced Martin as the Royal Society's London printer.

Flamsteed, John (1646–1719)

John Flamsteed's father was a prosperous business man. Between 1662 and 1669 Flamsteed studied astronomy on his own because poor health kept him from university. He began systematic astronomical observation in 1671, and began corresponding with Henry Oldenburg and John Collins. These two arranged for Flamsteed to meet Jonas Moore during a visit Flamsteed made to the Royal Society in London in 1670. Jonas Moore became his patron and persuaded Charles II to grant a warrant so that Jesus College Cambridge could award an M.A. to Flamsteed in 1674. In 1675 the King appointed Flamsteed his astronomical observer by Royal Warrant. From his salary of £100 he had to pay 10 taxes and also provide all his own instruments. In 1677 he was elected a fellow of the Royal Society. He worked for his entire life on his star charts for the northern hemisphere, consistently refusing to publish them on the grounds that the calculations were incomplete. They eventually appeared after his death.

Galilei, Galileo (1564–1642)

Son of a professional musician, Galileo studied medicine at the university of Pisa, but his real interests were always in mathematics and natural philosophy. After teaching mathematics privately in Florence and then at the University of Pisa, Galileo was appointed Professor of Mathematics at the University of Padua in 1592. In 1609 Galileo made a series of telescopes based on a Dutch prototype. In his *Starry Messenger*, published in 1610, Galileo announced he had seen four satellites orbiting Jupiter, using his telescope. He named them 'the Medicean stars', and shortly afterwards was appointed 'Mathematician and

Natural Philosopher' to the Medici family. He was examined by the Inquisition for maintaining that the earth moved in orbit round the sun (an official heresy), and refused to retract his published view to that effect in his *Dialogue* concerning two world systems (1632). He was condemned to house arrest and forbidden to publish. His *Discourses* on two new sciences was smuggled out of Italy and published in Leiden in 1638.

Gregory, David (1659–1708)

A nephew of James Gregory, he studied at the University of Aberdeen between 1671 and 1675, starting his university education at the age of twelve. He was appointed Professor of Mathematics at the University of Edinburgh at the age of twenty-four, where he taught Newton's theories for the first time. In 1691 Gregory left Scotland, and was elected Savilian Professor at Oxford (with Newton's strong backing) and a Fellow of the Royal Society. He also worked on optics, and experimented with making an achromatic telescope.

Gregory, James (1638–75)

James Gregory is credited with describing the first practical reflecting telescope, described in his 1663 *Optics*. He was appointed Professor at St Andrews in 1668 and became the first Professor of Mathematics at Edinburgh in 1674.

Halley, Edmond (1656–1742)

Halley assisted Flamsteed with astronomical observations from 1675, and became a member of the Royal Society at the age of 22. He went to St Helena in 1676, where he established the positions of about 350 southern hemisphere stars and observed a transit of Mercury. He proposed using transits of Mercury and Venus to determine the distance of the sun. Using his theory of cometary orbits he calculated that the comet of 1682 (now called

Halley's Comet) was periodic, and predicted that it would return in 76 years. He was appointed Savilian Professor of Geometry at Oxford in 1704. In 1720 he succeeded John Flamsteed as Astronomer Royal. Among his numerous other activities, Halley served as Deputy Controller of the Mint at Chester, and a sea captain on three scientific sea voyages in the late 1690s, mapped the English channel, and invented the first deep-sea diving suit.

Harvey, William (1578–1657)

Harvey was born into a well-to-do, landed family near Folkstone. After Cambridge, he travelled to Padua, where he studied medicine, graduating in 1602. He began to practise medicine in London in 1603, rising rapidly, helped by his marriage to the daughter of Elizabeth I's physician and later King James I's, Lancelot Browne. By 1607 Harvey had been admitted to the Royal College of Physicians. In 1615 he became Lumleian Lecturer there, a lifetime appointment. In 1616 he became physician to James I, eventually rising to become Principal Physician to Charles I, a position he continued to hold throughout the Civil War, right up to Charles's execution in 1649. Because of his close connection with the Crown, Harvey was fined 2000 pounds and banished from London for two years by the Commonwealth Government. He lived the remainder of his life in retirement. His great work on the circulation of the blood was based on large numbers of animal dissections, most of them conducted on animals killed by James and Charles during their frequent hunting sessions.

Hermann, Paul (1646–95)

Hermann was of German origin, but studied medicine at Leiden and Padua. He was appointed to the Dutch East India Company as a medical officer, and, on his way to Ceylon in 1672, called at the Cape, where he was the first known person to make a herbarium collection of local plants. Hermann returned to Leiden in 1680 where he became Professor of Botany and director of the botanical garden. He devoted himself to building up the Leiden garden with rare plants from the East Indies, America and the Cape. In 1682 he visited England, where he arranged to exchange plants with the Chelsea Physic Garden, and struck up a lasting friendship with William Sherard, with whom he exchanged plants and specimens. His *Paradisus Batavus* was published posthumously in 1698, edited by Sherard. Hermann's Cape and Ceylon herbaria were auctioned in 1711, when Petiver acquired a collection of Cape plants for Sloane.

Hevelius, Johann (1611–87)

Hevelius came from a family of wealthy brewers in Gdansk, and was a city councillor and Consul, active in public affairs. (Gdansk was a prosperous city, one of the four principal Hanseatic ports.) Hevelius studied law at Leiden and visited England and France as a young man. When he became interested in astronomy, he built an observatory on his own property. After his first wife died he married the sixteen-year-old daughter of a Gdansk merchant, who became his astronomical assistant in 1662. Colbert contributed to the running costs of Hevelius's observatory in the 1660s, in exchange for data. Hevelius was considered the most distinguished astronomer in Europe. In the tradition of Tycho Brahe, he observed with the naked eye, using a quadrant, a large sextant, and two pendulum clocks. His most important works were the *Selenographia* and *Cometographia*. Halley visited him in 1679.

Hire, Philippe de la (1640–1718)

French astronomer and mathematician, son of the Parisian painter Laurent de la Hyre. La Hire was elected to the Académie Royale des Sciences in 1678. In 1683 he was appointed to the chair of mathematics at the Collège Royale. Four years later he was appointed, in addition, to the chair of architecture at the Académie Royale. He worked on the extension of the meridian and the map of France for the Académie. He kept Huygens informed about the Académie after the latter had returned to The Hague in 1681.

Hobbes, Thomas (1588–1679)

Philosopher and keen amateur scientist (Aubrey reported that he had been amanuensis to Sir Francis Bacon in his youth), and author of Leviathan. Hobbes was hostile to the Royal Society because he considered it elitist and exclusive.

Hollar, Wenceslaus (1607–77)

Prolific Dutch engraver, who made several trips to the Tangier Mole and made official drawings and maps for the Navy Board.

Hooke, Robert (1635–1703)

Energetic inventor and experimentalist, and Curator of Experiments to the Royal Society. Renowned for his competitiveness, the speed of his ingenuity and his capacity for quarrelling with other scientists about priority. Provided specialist surveying services to the team rebuilding London after the Great Fire of 1666, and assisted Sir Christopher Wren in designing the new St Paul's Cathedral. No portrait appears to survive.

Howard, Henry Duke of Norfolk (1628–84)

Grandson of the diplomat, bibliophile and collector of antiquities, Thomas Arundel, and keen scientific amateur. Appointed Earl Marshal in 1682 (titular head of the College of Arms). The Arundels were traditionally Catholics, but Thomas Arundel converted to Protestantism. One of Henry's brothers, however, was executed for involvement in the Popish Plot in 1680.

Huygens, Christiaan (1629–95)

Dutch mathematician and astronomer who became one of the
leading European scientific figures of his day. He was appointed
by Louis XIV as a founding member of the Académie des
Sciences in Paris on a pension of 5000 livres in 1666. In spite of
his Protestantism, he continued there (apart from the occasions
when bouts of illness necessitated his return to The Hague)
until the early 1680s. He visited England on several occasions,
was a foreign member of the Royal Society, and corresponded on
a regular basis with a number of other members, including Moray
and Oldenburg.

Huygens, Constantijn (junior) (1628–97)

Followed his father into diplomacy in the Dutch Republic. Kept
up a vigorous correspondence with his brother Christiaan during
the latter's tenure as a founding member of the Académie des
Sciences in Paris. Came to England with William of Orange in
1688.

Huygens, Constantijn (senior) (1596–1687)

Senior diplomat in the Dutch Republic, with a keen amateur
interest in art and science. He encouraged his son Christiaan to
become a mathematician and scientist.

Kepler, Johannes (1571–1630)

Kepler attended the University of Tübingen, where his mathematical ability was noticed by his astronomy teacher Michael
Maestlin. Kepler originally intended to become a priest, but
instead took up a post teaching mathematics at Graz. In his
Mysterium Cosmographicum (1596), Kepler argued for the truth of
the Copernican system. In 1600, Kepler went to Prague, as assistant to Tycho Brahe. When Brahe died in 1601, Kepler continued
his work and used his observations to calculate planetary orbits

with unparalleled accuracy. Kepler showed that a planet moves round the Sun in an elliptical orbit which has the Sun in one of its two foci. He also showed that a line joining the planet to the Sun sweeps out equal areas in equal times as the planet describes its orbit. Kepler also proved that sight was triggered by the reception of light rays in the eye (1604), and wrote on the optics of the telescope, introducing the design using two convex lenses (1611). His *Rudolphine Tables* (1627), based on Tycho's observations and Kepler's laws, proved to be accurate over a long time-scale, and were widely used by navigators. Their success did much to gain general acceptance for the heliocentric theory.

Kirsch, Gottfried (1640?–1710)

German astronomer who trained with Hevelius at Gdansk, and in 1700 became Astronomer at the Royal Academy of Sciences in Berlin. He married Maria Winkelmann; after his death in 1710 she was refused the post of Astronomer in his place.

Leeuwenhoek, Antoni van (1632–1723)

Civil servant in Delft in the Dutch Republic, and amateur microscopist. From 1673 he corresponded regularly with the Royal Society, via Oldenburg, and provided a succession of observations of the minute structure of natural phenomena, from the interior structure of plants, to the protozoa in pepper-water, and male sperm in semen.

Leibniz, Gottfried Wilhelm (1646–1716)

Born Baron Leibniz in Leipzig, the son of a philosophy professor, Leibniz showed an early ability in mathematics, but also had interests in languages, literature, law and philosophy. He became a diplomat, and in 1672 was sent by the Elector of Mainz to Paris to try to dissuade Louis XIV from attacking German territories. During this period he developed the differential calculus (although Newton maintained he had stolen the idea from him). He remained in Paris for four years, studying mathematics with Huygens. He visited England and became an overseas member of the Royal Society. After 1676 he spent the rest of his life in Germany. Leibniz founded the Berlin Academy in 1700 and was its first president.

Lister, Martin (c. 1638–1712)

Lister was a medical doctor, and collector of insects and shells. He compiled and published a magnificent *History of Shells*, which was lavishly illustrated, thanks to the talent and painstaking efforts of Lister's daughters Susanna and Anna, who drew and engraved over 1,000 shells for it. He was involved in the medical circle, including Wren, which carried out dissections and experimented with intravenous injection.

Locke, John (1632–1704)

Physician and philosopher, who was also greatly interested in medical materials, and collected a substantial herbarium.

Malpighi, Marcello (1628–94)

Malpighi was born in Crevalcore in Italy. He obtained a degree in medicine and philosophy at the University of Bologna in 1653. As Professor of Theoretical Medicine at the University of Pisa, he began his microscopic observations and became strongly critical of the assumptions held about physiology and medicine. In 1659 he returned to Bologna and in 1661 made his most important discovery, describing the network of pulmonary capillaries that connect the small veins to the small arteries, thus completing the chain of circulation postulated by the English physician William Harvey. His work was taken up by Oldenburg, who presented it to the Royal Society, which published his major works in London.

Maskelyne, Nevil (1732–1811)

Maskelyne studied mathematics at Cambridge, graduating seventh in his year in 1754. He was ordained a minister in 1755, and became a Fellow of Trinity College Cambridge in 1756. In 1758 he became a member of the Royal Society, which in 1761 sent Maskelyne to the island of St Helena (in the footsteps of Halley) to observe a transit of Venus. During the voyage he experimented with the lunar position method of determining longitude. Maskelyne returned to Chipping Barnet in 1761, where he was a curate, and published his lunar distance method for determining longitude in *The British Mariner's Guide* (1763). In 1764 he went on a voyage to Barbados to carry out trials of Harrison's longitude chronometer. Soon after his return, in 1765, he was appointed Astronomer Royal.

Mason, Charles (1730–87)

Mason was James Bradley's assistant at the Greenwich Observatory from 1756 to 1760. He and Jeremiah Dixon were chosen by the Royal Society to observe the transit of Venus of 6 June 1761. They did so successfully at the Cape of Good Hope, having been prevented by a variety of disasters from reaching their original destination of Sumatra. On the way home, Mason co-operated with Nevil Maskelyne in collecting tidal data on St Helena (where Maskelyne had observed the transit of Venus). Mason and Dixon were next hired by Lord Baltimore and Mr Penn to settle the boundary between Maryland and Pennsylvania. They also measured an arc of the meridian there for the Royal Society in 1764. Mason was employed by the Royal Society during six months in 1769 on an astronomical mission to Ireland, where he observed a second transit of Venus, a partial solar eclipse, eclipses of Jupiter's satellites, and in August and September the famous comet that signalised the birth year of Napoleon. He returned to America, and died in Philadelphia in 1787.

Maupertuis, Pierre Louis Moreau de (1698–1759)

Maupertuis became a member of the Académie Royale des Sciences in Paris in 1731. In 1732 he introduced Newton's theory of gravitation to France. He was a member of an expedition to Lapland in 1736 that set out to measure the length of a degree along the meridian. Maupertuis's measurements verified Newton's claim that the Earth would be an oblate spheroid. As a result of the success of this expedition, he was invited to Germany by Frederick the Great. He became a member of the Berlin Academy of Sciences in 1741 and, four years later, became its president. He held the post of president for eight years.

Merian, Maria Sybilla (1647–1717)

Born into a family of well-educated craftsmen, Merian was the daughter of the famous artist and publisher Matthäus Merian the Elder, in Frankfurt am Main. She married another artist, Johann Andreas Graff, in 1665. They moved to Nurenburg in 1670, where Merian pursued a career as a professional artist. In 1681 Merian moved back to Frankfurt with her two daughters. In 1685 she left her husband and joined the Labadist community in Wieuwerd, effectively ending her marriage. When the sect collapsed she returned to Amsterdam in 1691 and supported herself and her daughters with her drawings of plants and insects. She went to Surinam in 1699, funding her journey by selling her large collection of pictures and preserved specimens. After two years the climate was too much for her and she returned to Amsterdam in September 1701. She was employed to provide the replacement plates for Rumphius's vast work on the flora and fauna of Amboina, which provided her with something to live on – otherwise she supported herself thenceforth by selling her Surinam drawings (a printed volume of these was also published), and importing and selling Surinam specimens.

Mersenne, Marin (1588–1648)

Marin Mersenne is best known for his prolific correspondence with eminent philosophers and scientists of his day. From 1604 he spent five years in the Jesuit College at La Fleche. From 1609 to 1611 he studied theology at the Sorbonne. He joined the religious order of the Minims in 1611. He taught philosophy at the Minim convent at Nevers from 1614 to 1618. In 1619 he returned again to Paris, where his monastic cell became a meeting place for Fermat, Pascal, Gassendi, Roberval, Beaugrand and others who later became the core of the Académie Royale des Sciences. Mersenne was a good mathematician, whose correspondence with other eminent mathematicians played a major role in communicating mathematical knowledge throughout Europe at a time when there were few scientific journals.

Moore, Jonas (1617–79)

In 1647 Moore became, for a short time, mathematics tutor to the Duke of York, the future James II. Moore then went to London where he taught mathematics. In 1649 he was appointed as chief surveyor to work on the draining of the Fens, a natural region of about 40,100 square kilometres of reclaimed marshland in eastern England between Lincoln and Cambridge. Around the same time, he published a mathematical textbook (1650). In 1663 Moore was sent to Tangier to conduct a survey and to report on its fortifications, becoming involved in the ambitious project to build a massive harbour wall or mole. He received a knighthood in 1669 and was appointed to high office as Surveyor-General of the Ordnance. The Royal Observatory at Greenwich was established under his direct patronage, and he supported Flamsteed, the first Astronomer Royal, until his own death. As a further injection of scientific expertise into the military matters which were their first concern, Moore and Pepys (who was on the Navy Board) founded the Royal Mathematical School within Christ's Hospital to train boys in navigation techniques.

Moray, Robert Sir (d. 1672)

Influential royalist courtier, and go-between between the court and the Royal Society (of which he was a founder member, and in which he took a lifelong interest), Moray was an energetic correspondent with scientists internationally. He was Master of the Scottish Lodge of Freemasons.

Newton, Isaac (1642–1727)

Newton never knew his father, who died before he was born. When his stepfather died in 1656, Newton's mother removed him from grammar school in Grantham where he had shown little promise in academic work. An uncle fortunately decided that he should go to university, and he entered Trinity College, Cambridge, in June 1661, ostensibly to study law, but soon began to show exceptional ability in mathematics. In the summer of 1665 plague closed the university and he had to return to Lincolnshire. While at home he made a number of important advances in mathematics. In 1669 Barrow resigned the Lucasian Chair, recommending that the twenty-seven-year-old Newton be appointed in his place. Newton was elected a Fellow of the Royal Society in 1672 on the basis of a reflecting telescope he had designed. Newton's greatest achievement was his work in physics and celestial mechanics, which culminated in the theory of universal gravitation. By 1666 Newton had early versions of his three laws of motion. He had also discovered the law giving the centrifugal force on a body moving uniformly in a circular path. From his law of centrifugal force and Kepler's third law of planetary motion, Newton deduced the inverse square law. In 1684 Halley consulted Newton about the orbits of comets, and subsequently persuaded Newton to write a full treatment of his new physics and its application to astronomy in his *Principia* (1687). After suffering some kind of nervous collapse in 1693, Newton took up a government position in London. He became Warden of the Royal Mint (1696) and Master (1699). In 1703 he was elected President of the Royal Society and was re-elected each year until his death. He was knighted in 1708 by Queen Anne, the first scientist to be so honoured for his work.

Oldenburg, Henry (1615?–77)

Born in Bremen in modern Germany, Oldenburg was the official representative of Bremen to Cromwell's Commonwealth Government. After the Restoration he remained in England, becoming a founding member of the Royal Society, and its first Secretary. His information-gathering via prolific correspondence with scientists and thinkers around Europe contributed significantly to advances in the new science in England. He was once incarcerated briefly as an intelligencer (spy) – a charge which is not implausible, given the vigorous nature of his transmission of information of all sorts to French and Dutch colleagues, even during times of conflict between these countries and England.

Papin, Denis (1647–1712)

Papin graduated with a medical degree from the University of Angers in 1669. He assisted Huygens with air pump experiments from 1671 to 1674, during which time he lived in Huygens's apartments in the Royal Library in Paris. In 1675 he went to London to work with Boyle. He remained in this post until 1679 when he became Hooke's assistant at the Royal Society. He was elected a Fellow of the Royal Society in 1680. In 1681 Papin left for Italy where he was director of experiments at the *Accademia publicca di scienze* in Venice until 1684. Because Papin was a Calvinist, born into a Huguenot family, he could not return to France after the Edict of Nantes, which had granted religious liberty to the Huguenots, was revoked by Louis XIV in 1685. In 1684 he returned to London to work with the Royal Society again. In 1687 Papin left England and went to Hesse-Kassel where he was appointed Professor of Mathematics at the University of Marburg, leaving the post in 1696. In 1707 he returned to England where he died in obscurity. He is credited with inventing the pressure cooker and the steam engine.

Pepys, Samuel (1633–1703)

Pepys wrote his diary during the period 1660 to 1669, and his records and comments form a vital part of our knowledge of the day-to-day life of the period. He held a variety of roles on the Navy Board, and was closely involved with the building of the Tangier Mole, and then its destruction.

Perrault, Claude (1613–88)

One of the leading botanists in the Paris Académie Royale des Sciences, to whom Colbert allocated the largest annual pension after Cassini and Huygens. A physician and an architect, he designed several royal buildings including the Paris Observatoire. Perrault directed the Académie's acclaimed Histoire des animaux. Christiaan Huygens was a personal friend.

Petiver, James (1663–1718)

Prominent London businessman, and keen collector of rarities. He later acted as Sloane's agent for purchasing, travelling abroad on his behalf to purchase the Dutch botanist Hermann's collection after his death. Petiver corresponded with Maria Sybilla Merian, and bought specimens from her. Sloane bought Petiver's own collection from his sister for four thousand pounds, after Petiver's death.

Picard, Jean (1620–82)

Jean Picard studied at the Jesuit college at La Flèche. In 1655 Picard became Professor of Astronomy at the Collège de France in Paris. He became a member of the Académie Royale des Sciences in 1666, just after its foundation, and from then on devoted himself to working for the Académie. In 1667 Picard added microscopic sights to the quadrant to improve its use in surveying and navigation. He measured the length of the arc of the meridian as part of the French mapping project. In 1673 Picard moved to the Paris Observatoire where he collaborated with Cassini, Römer and, slightly later, with La Hire. Picard was one of the first to apply scientific methods to the making of maps. He produced a map of the Paris region, then went on to join a project to map France.

Plot, Robert (1640–96)

In the 1670s Plot undertook a project for studying the natural history of England which materialised as his *Natural History of Oxfordshire* (1677) and *Natural History of Staffordshire* (1686). He was appointed first Keeper of the Ashmolean in 1683 and Professor of Chemistry in the same year.

Povey, Thomas (c. 1615–c. 1702)

Courtier and Administrator. Master of Requests to Charles II from 1662 till 1685. Treasurer to the Duke of York (later James II) and active member of the Royal Society. Povey was well known for his lavish tastes, and was a considerable collector and patron of fine art. He was a friend of Ashmole's, who presented him with a copy of his *Institution of the most Noble Order of the Garter*.

Power, Henry (1623–88)

Physician and microscopist, who published an unillustrated book on early microscopy.

Ray, John (1627–1705)

Botanist and taxonomist, protégé of Dr John Wilkins. Ray was the son of the blacksmith in a village near Braintree. He went to Trinity College, Cambridge, where he became a Fellow in 1649, and began to study botany. In 1662 he lost his fellowship, because of his religious nonconformity. Between 1663 and 1666 he travelled widely in the Netherlands, Germany, and Italy, with his well-to-do younger botanical colleague and friend Francis Willoughby. He became a member of the Royal Society in 1667. Willoughby died in July 1672, and his mentor Wilkins died in November the same year. Ray saw all Willoughby's work through the presses. His health was poor, and apart from brief excursions to London he was confined to his home for the rest of his life. His botanical works were compiled using specimens sent to him by friends, particularly Sloane, who also helped him financially, and prescribed for his unpleasant illnesses. The first two volumes of his *Historia plantarum generalis* came out in 1686 and 1688 with a supplement in 1704.

Reeve, Richard (fl. 1640–80)

London instrument maker, whose telescopes and microscopes had a worldwide reputation for accuracy. Hooke worked with him in a technical advisory capacity.

Rhijne, Willem Ten (1647–1700)

Born in the Netherlands, Ten Rhijne received his medical training at Franeker and Leiden. He was appointed as a physician in the Dutch East India Company and held responsible posts in Java and Japan. On his way to Java, he called at the Cape in 1673, and made sketches and collections of plants on and around Table Mountain and at Saldanha Bay. He assisted van Reede tot Drakenstein with the Latin descriptions for his *Hortus Malabaricus,* and checked the copy of Rumphius's *Herbarium Amboinense.*

Richer, Jean (1630–96)

Richer became a member of the Académie Royale des Sciences in 1666 with the title of 'astronomer'. By 1670, however, he had been given the title 'mathematician'. He spent the rest of his life working for the Académie. In 1670 Richer was sent to La Rochelle to measure the heights of the tides there at both the spring and vernal equinoxes. Also in 1670 he set out on a voyage to French Canada with the task of testing two clocks made by Huygens (the clocks stopped during a storm). In 1671 Richer was sent on an expedition to Cayenne, in French Guyana, by Cassini, the Royal Astronomer. His first task there was to measure the parallax of Mars and the observations were to be compared with those taken at other sites to compute the distance to the planet. Richer's also observed how a pendulum clock performed at Cayenne (it ran slow). Newton and Huygens used Richer's data to show that the Earth is an oblate sphere. In 1673 Richer returned to Paris where he was given the title of 'Royal Engineer' and undertook work on fortifications.

Ruysch, Frederik (1638–1731)

Amsterdam anatomist and microscopist, renowned for his invention of a method for injecting the fine vessels in cadavers with tinted wax for display purposes. Originally a professor of anatomy, he became the first Professor of Botany associated with the Amsterdam Botanical Garden after its foundation in 1682.

Sloane, Hans (1660–1753)

Sloane was of Scots descent, whose family moved to Ireland under James I's settlement of that 'colony'. He was born at Killyleagh, County Down. His father was the agent for James Hamilton, later Earl of Clanbrassill. He trained in medicine in London, coming heavily under the influence of Ray (who was thirty years older than him). He went to France in 1683 with

Tancred Robinson, where they pursued an intensive programme of training in the Royal Garden of Plants with Tournefort. He moved on to Montpellier. In 1684 he returned to London, settled in Fleet Street, and quickly gained a reputation as a physician, gaining the support of the leading London doctor Thomas Sydenham, and becoming a member of the Royal College of Physicians in 1687. He also began his collections of botany and rarities. In autumn 1687 he went to Jamaica as physician to the new Governor, Lord Albemarle, returning in 1689 (just after the arrival of William III and Mary) after Albemarle's death. From marketing the chocolate and other 'medical' materials he found in Jamaica he made a substantial fortune that he ploughed into his collections. Under his will he established these as the foundation for the British Museum. In later life he was a leading figure in the Royal Society, and its president from 1727 for fourteen years, standing down at the age of eighty-one.

Sluze, René François de (1622–85)

René de Sluze studied at the University of Louvain from 1638 to 1642, then went to Rome where he received a law degree from the University of Rome in 1643. He became a canon in the Catholic church in 1650. By 1659 he was a member of the privy council of the Bishop of Liège; he became abbot of Amay in 1666. De Sluze wrote a large number of books on mathematics. He corresponded with mathematicians in England and France. He was elected an overseas Fellow of the Royal Society in London in 1674.

Tradescant, John (senior) (1577–1638)

Gardener to Sir Robert Cecil, the Duke of Buckingham, Charles I and James I, Tradescant transformed English gardens, and embarked on an ambitious personal programme of growing imported rarities in his aristocratic gardens. He travelled as far as

Russia himself, bringing back numerous previously unencountered species. The Tradescant Cherry appears in contemporary recipes in place of the morello cherry. He is credited with introducing the Horse Chestnut, whose fruit he had been told were edible. The avenues of these trees on the English landscape are a lasting memorial to him. His large collection of curiosities, which he collected alongside his plants and trees, was on view to the public at his home in Lambeth, known as the 'Ark' because of its diversity. After his death, his son continued the collection. Ashmole acquired it, under somewhat murky circumstances from his widow.

Tschirnhaus, Ehrenfried Walter von (1651–1708)

Born in Germany, Tschirnhaus visited England in 1674 where he met Wallis in Oxford and Collins in London. He also visited Leiden and then Paris, where he remained for a while after meeting Leibniz and Huygens. Tschirnhaus's main aim in life for many years was to obtain a paid position at the Académie Royale des Sciences in Paris. He was elected a member in 1682 but without remuneration. Tschirnhaus was a scientist, and among other things, he experimented making porcelain from clay mixed with fusible rock in the 1680s. A factory at Meissen started production of his porcelain in 1710 and Tschirnhaus's porcelain was sold at the Leipzig Fair in 1713. Tschirnhaus, however, refused to divulge the secret of his porcelain manufacture, and ended his life heavily in debt.

Wharton, Thomas (1614–73)

Eminent physician and anatomist, author of *Adrenographia* (1656) and a prominent member of the London College of Physicians before and after the Restoration. 'Most beloved friend' of Ashmole, according to Anthony Wood. He gave Ashmole medical advice and Ashmole reciprocated with astrological information. They collaborated in cataloguing the Tradescant Rarities.

Wilkins, John, Bishop of Chester (1614–1672)

Keen scientific enthusiast and close friend of Hooke and Ray. First gained advancement during the Commonwealth period. Married Oliver Cromwell's sister. Dismissed from the Mastership of Wadham College, Oxford at the Restoration of Charles II in 1660, but gradually re-established his position. Founder member of the Royal Society.

Winkelmann, Maria Margaretha (1670–1720)

Born near Leipzig and privately educated by her father. Became unofficial apprentice to astronomer Christoph Arnold. Married astronomer Gottfried Kirsch, who had trained with Hevelius in Gdansk. In 1700 they moved to Berlin, where Kirsch became astronomer at the Royal Academy of Sciences; Winkelmann became an Academy calendar-maker. She published three astronomical pamphlets, but in 1711, after the death of her husband, she failed to obtain his post as astronomer. Thereafter she continued her astronomical work privately.

Wren, Christopher (1632–1723)

Wren is best known as an architect, but his interests also included geometry and astronomy. He entered Wadham College, Oxford in 1649, graduating with an MA in 1653. In 1657 he became Professor of Astronomy at Gresham College, London. Wren became Savile Professor of Astronomy at Oxford in 1661 and held this post until 1673. In 1664, he designed the Sheldonian Theatre, Oxford, and in 1665, buildings for Trinity College, Oxford. Following the Great Fire of London in 1666 he became one of the surveying team appointed to rebuild the city, and designed and built the new St Paul's Cathedral. Wren was also an able mathematician, highly regarded by Newton. Wren was a founder member of the Royal Society in London, and president from 1680 to 1682.

NOTES

INTRODUCTION

1 I. Wilmut *et al*, 'Viable offspring derived from fetal and adult mammalian cells', *Nature*, 385 (1997), 810–13.

2 N. Rosenthal and S. Fraquelli (eds), *Sensation: Young British Artists from the Saatchi Collection* (London: Royal Academy of Arts in association with Thames and Hudson, 1997).

3 Bronowski, *Science and Human Values*.

4 On the anthropology of science, see Latour, *Laboratory Life*.

5 For the opposed view that science is a distinct practice, utterly unlike 'art' or even 'common sense', see Dawkins and Wolpert.

6 See, however, the wonderful opening sentence of Steve Shapin's recent book, *The Scientific Revolution*: 'There was no such thing as the Scientific Revolution, and this is a book about it.'

7 Jardine, *Worldly Goods*.

CHAPTER I: SIGNS OF THE TIMES

1 Cook, *Halley*, 109.

2 Westfall, *Never at Rest*, 391.

3 See Andrewes, *The Quest for Longitude*; Sobel, *Longitude*, 31.

4 See Howse, *Greenwich Time*, passim.

5 Stroup, *A Company of Scientists*, 43–4.

6 On Moore see Willmoth, *Sir Jonas Moore*.

7   Willmoth, *Sir Jonas Moore*, 165–95.

8   Willmoth, 'Mathematical sciences and military technology', 129–30.

9   Cook, *Halley*, 53–4.

10 Howse, *Greenwich Time*, passim.

11 Howse, 'Newton, Halley, and the Royal Observatory', 16.

12 Ibid., 22–3.

13 Westfall, *Never at Rest*, 694–6; Howse, 'Newton, Halley and the Royal Observatory', 22–3.

14 Westfall (ed.) Newton's *Principia* book 3.

15 Westfall, *Never at Rest*, 393.

16 Schaffer, 'Newton's comets and the transformation of astrology', 225.

17 cit. Genuth, *Comets*, 133; Westfall, *Never at Rest*, 60.

18 Westfall, *Never at Rest*, 65.

19 Ibid. 104.

20 Arthur Koestler, *The Sleepwalkers*, 234.

21 Bennett, *The Mathematical Science of Christopher Wren*, 65–70.

22 Schaffer, 'Newton's comets', 226.

23 On Halley's relations with Flamsteed see Cook, *Halley*, 380–7.

24 A. Cook, *Halley*, 62.

25 Waters, 'Captain Edmond Halley, F.R.S., Royal Navy, and the practice of navigation', 177–80.

26 Birch, *History of the Royal Society* 3, 362; Cook, *Halley*, 74–5.

27 Ibid. 3, 387.

28 Taylor, 'Robert Hooke and the cartographical projects', 540.

29 Thrower, 'The Royal patrons of Edmond Halley, with special reference to his maps', 204.

30 Whitfield, *The Mapping of the Heavens*, 97.

31 Biagoli, *Galileo Courtier*, 131–57.

32 Whitfield, *The Mapping of the Heavens*, 88.

33 Cook, *Halley*, 91.

34 Stroup, *A Company of Scientists*, 31.

35 Fara, *Sympathetic Attractions*, 109.

36 Thrower, *Maps and Civilization*, 97.

37 Waters, 'Halley and the practice of navigation', 183–4.

38 D. W. Hughes, 'Edmond Halley: his interest in comets', 338.

39 Cook, *Halley*, 152–3.

40 E. F. MacPike (ed.), *Correspondence and Papers of Edmond Halley* (London: Taylor and Francis, 1937), 91.

41 Ibid., 92.

42 Ibid., 238.

43 Martins, 'Huygens's reaction to Newton's theory of gravity', 204.

44 Cook, *Halley*, 161.

45 Jacob, *Newtonians and the English Revolution*, 135–6.

46 Genuth, *Comets*, 165.

47 Ibid., 151.

Chapter 2: Scale Models

1   cit. J. T. Harwood, *Robert Hooke: New Studies*, 120.

2   Pepys, *Diary* 5, 240.

3   Pepys, *Diary* 5, 241; cit. Harwood, *Robert Hooke: New Studies*, 145.

4   Hooke, *Micrographia*, preface.

5   See R. H. Nuttall, 'That Curious Curiosity: the Scotoscope', *NRRS* 42 (1988), 133–8, cit. Simpson, 'Robert Hooke and practical optics', 46.

6   H. W. Jones (ed. and transl.), *Thomas Hobbes: Thomas White's De Mundo Examined* (London: Bradford University Press, 1976), 119–20.

7   Hunter and Schaffer, *Robert Hooke: New Studies*, 37.

8   Ibid., 38.

9   Gingerich, 'How astronomers finally captured Mercury', 38–9.

10 Letter from Hooke to Boyle, 21 October 1664, Gunther 6, 206.

11 Pepys, *Diary* 7, 254.

12 Ibid.

13 Ibid., 5, 240.

14 Letter from Hooke to Boyle, 3 July 1663, Gunther 6, 139.

15 Shapin and Schaffer, *Leviathan and the Air-Pump*, 128.

16 See chapter 3, below.

17 Shapin and Schaffer, *Leviathan and the Air-Pump*; Frank, *Harvey and the Oxford Physiologists*.

18 Shapin and Schaffer, *Leviathan and the Air-Pump*, 231; Frank, *Harvey and the Oxford Physiologists*, 130.

19 Evelyn, *Diary*, 132.

20 Shapin and Schaffer, *Leviathan and the Air-Pump*, 30–31.

21 cit. Drake, *Restless Genius*, 21.

22 Stroup, *A Company of Scientists*, 160–65.

23 Schiebinger, 'Maria Winkelmann', 319.

24 'Espinasse, *Hooke*, 51–2.

25 S. Shapin, *A Social History of Truth*, 356.

26 See A. D. C. Simpson, 'Robert Hook and practical optics'.

27 Westfall, *Never at Rest*, 234.

28 Ibid., 237.

29 Ibid., 660.

30 Ibid., 81.

31 Ibid., epigram.

32 Hooke, *Micrographia*, fol. a2v.

33 Bennett, *The Mathematical Science of Christopher Wren*, 40, 73.

34 Wren, *Parentalia*, 211.

35 cit. *Robert Hooke: New Studies*, 129.

36 Ibid (Boyle, *Works* v, 534).

37 Birch, *History of the Royal Society* I, 466 (Boyle, *Works*, v, 535).

38 Birch, *History of the Royal Society* I, 471.

39 See Birch, *History of the Royal Society* I, 456 for the instruction to Hooke.

40 Hunter, *Science and the Shape of Orthodoxy*, 63–5.

41 F. Willmoth, *Sir Jonas Moore*, 143.

42 H. F. Hutchinson, *Sir Christopher Wren: A Biography*.

43 Evelyn, *Diary*, 171.

44 F. Willmoth, *Sir Jonas Moore*, 63; Smith, *History of the Modern British Isles*, 217.

45 Evelyn, *Diary*, 173–4.

46 *Sir Christopher Wren: An Exhibition Selected by Kerry Downes*, 72.

47 This is the period covered by the first Hooke diary. See Robinson and Adams (eds), *The Diary of Robert Hooke*.

48 These visits and this on-site involvement continue in the later diary, printed in Gunther, *Early Science in Oxford* 10.

49 'Espinasse, *Hooke*, 104.

50 Drake, *Restless Genius*, 131.

51 Load-bearing arches: Hooke, *Diary*, 8 December, 15 December, 12 January, 19 January 1671.

52 Gunther 8, 151.

53 Ibid., 131.

54 On the Hooke/Wren friendship see Frank, *Harvey and the Oxford Physiologists*, especially chapter 3.

55 Bennett, *The Mathematical Science of Christopher Wren*, 41–2.

56 Hooke, *Diary*, 358.

57 Ibid., 359.

58 Birch, *History of the Royal Society* 3, 409–10.

59 Ibid., 388.

60 cit. Simpson, 'Robert Hooke and practical optics', 39–41. See Hunter, 'A "College" for the Royal Society', 173.

61 Royal Society MS Council Minutes 2 (copy), p. 169; cit. Bennett, 42.
62 Bennett, *The Mathematical Science of Christopher Wren*, 42.
63 Evelyn, *Diary*, 305.
64 Ibid., 143.
65 cit. Shapiro, *John Wilkins*, 236.
66 Gunther 6, 191.
67 Ibid., 6, 195.
68 See Schaffer and Shapin, *Boyle and the Air Pump*.
69 Gunther 6, 222–3. For the engravings themselves, see Birch, *Works* 2.
70 Gunther 6, 223.
71 Ibid.
72 Gunther 6, 226.
73 See, for example, Dora Thornton, *The Scholar in his Study: Ownership and Experience in Renaissance Italy* (New Haven: Yale University Press, 1997)

CHAPTER 3: SUBTLE ANATOMY

1   Schierbeek, *Measuring the Invisible World*, 60; Wilson, *The Invisible World*, 88–90.
2   Schierbeek, *Measuring the Invisible World*, 30.
3   On Oldenburg see Hunter, 'Promoting the new science', and Oldenburg, *Correspondence* I, Introduction.
4   Schierbeek, *Measuring the Invisible World*, 63–4.
5   Birch, *History of the Royal Society* 3, 352.
6   Schierbeek, *Measuring the Invisible World*, 69.
7   Ibid., 35.
8   Ruestow, *The Microscope in the Dutch Republic*, 6–24.
9   S. Schaffer, 'Regeneration: the body of natural philosophers in Restoration England', 93.
10 Ruestow, *The Microscope in the Dutch Republic*, 24–5.
11 Ibid., 26.
12 Hooke, *Micrographia*, fol. f2v.
13 Ruestow, *The Microscope in the Dutch Republic*, 155–6.
14 Taken from the 1681 English edition of *The Anatomy of the Brain* (1681) (reprint, USV Pharmaceutical Corporation, Tuckahoe, New York, 1971), 4.
15 Alpers, *The Art of Describing*, 6–7.
16 Ruestow, *The Microscope in the Dutch Republic*, 106.
17 Alpers, *The Art of Describing*, 51; *Reliquiae Wottonianae, or, A Collection of Lives, Letters, Poems; with Characters of sundry personages* (London, 1651), 413–14.

18 C. Brusati, *Artifice and Illusion*, 71.

19 Ibid., 92–3.

20 Hooke, *Micrographia*, 186.

21 Willis and Guyton (transl. and ed.), *The Works of William Harvey*, 76.

22 Frank, *Harvey and the Oxford Physiologists*, 1.

23 J. Spedding, R.L. Ellis and D.D. Heath, *The Works of Francis Bacon*, 7 (London: Longman et al., 1857–9), 3, 159.

24 Harvey, *Works*, 7.

25 A. Cunningham, *The Anatomical Renaissance*, 183.

26 Harvey, *Works*, 19.

27 Ibid., 31.

28 On Harvey's insistence on ocular confirmation see A. Wear, 'William Harvey and the "way of the anatomists"'.

29 J. Sawday, *The Body Emblazoned*, 240.

30 Evelyn, *Diary*, 132.

31 'Espinasse, *Hooke*, 52.

32 Evelyn, *Diary*, 184.

33 Wren, *Parentalia*, 229–30.

34 Pepys, *Diary* 5, 151.

35 Birch, *History of the Royal Society* 2, 186.

36 Letter from Oldenburg to Boyle, 13 September 1667, Birch, *Boyle Works* 6, 238–9.

37 S. Schaffer, 'Regeneration', in *Science Incarnate*, 100–101.

38 Birch, *History of the Royal Society* 2, 216.

39 Evelyn, *Diary*, 185.

40 Schaffer, 'Regeneration', in *Science Incarnate*, 102–3.

41 Ibid., 103.

42 Birch, *Boyle Works* 6, 245.

43 Ibid., 6, 257.

44 Ibid., 6, 260.

45 See Oldenburg *Correspondence* 4, 237–43; Frank.

46 Evelyn, *Diary*, 183.

47 Wilson, *The Invisible World*, 35.

48 Ibid., 124–5.

49 Ruiestow, *The Microscope in the Dutch Republic*, 105.

50 Ibid., 86–7.

51 Oldenburg, *Correspondence* 8, 617–9.

52 Ibid., 9, 41–2.

53 Ibid., 9, 367–9.

54 Birch, *History of the Royal Society*, 3, 71.

CHAPTER 4: RUNNING LIKE CLOCKWORK

1 Landes, *Revolution in Time*, 105 and *passim*.
2 Ibid., 104.
3 Ibid., 105.
4 See Andrewes, 'Finding local time at sea'.
5 Landes, *Revolution in Time*, 106. For clear accounts of the chronometer and astronomical methods see Andrewes (ed.), *The Quest for Longitude*; for the story of the triumph of the chronometer method see Sobel, *Longitude*.
6 A. van Helden, 'Longitude and the satellites of Jupiter', in Andrewes (ed.), *The Quest for Longitude*, 93–96.
7 cit. van Helden, 'Longitude and the satellites of Jupiter', 94.
8 D. Howse, *Nevil Maskelyne: The Seaman's Astronomer*, 27–8.
9 Howse, *Maskelyne*, 108–9.
10 Landes, *Revolution in Time*, 125.
11 Oldenburg *Correspondence* 11, 186 (translation from the French taken from the version published by Oldenburg in the *Philosophical Transactions* of the Royal Society in November 1676).
12 On 17 February [= 27 February; London dating ran ten days behind Paris]. Hooke, *Diary*, 147.
13 Hooke, *Diary*, 148.
14 Leopold, 'The longitude timekeepers of Christiaan Huygens', in Andrewes (ed.), *The Quest for Longitude*, 103–6.
15 Andrewes, 'Finding local time at sea', 396.
16 Leopold, 'The longitude timekeepers of Christiaan Huygens', 106.
17 Gunther 6, 238–9.
18 Wright, 'Robert Hooke's longitude timekeeper', 76.
19 See Hooke's various 'postscript' and 'appendix' vindications of his priority on the hair-spring movement in *Helioscopes*, *Lampas* and *Of Spring*, all reprinted in Gunter 8.
20 Letter from Hooke to Boyle, August 1665, in Gunter 6, 250.
21 Shapiro, *John Wilkins*, 199–200.
22 Gunther 6, 249–50
23 See Andrewes, 'Finding local time', 400–1 for Hooke's reflecting quadrant.
24 The most plausible reconstruction of these events is Wright, 'Robert Hooke's longitude timekeeper', 63–118; see also Iliffe, 'Privacy, property and priority', and Espinasse, *Hooke*, 64.

25 See below, chapter 8.

26 Oldenburg, *Correspondence* 9, 225–6.

27 Hooke, *Diary*, 157.

28 cit. Iliffe, 'Piracy, property and priority', 46.

29 Hooke, *Diary*, 166.

30 Willmoth, *Sir Jonas Moore*, 61–2.

31 cit. Ibid., 167–8.

32 cit. Ibid., *Sir Jonas Moore*, 164.

33 Hooke, *A Description of Helioscopes, and some other Instruments* (London: John Martyn, Printer to the Royal Society, 1676), reprinted in Gunter 8, 142. For the fact that Halley and Streete were also at Moore's for this eclipse see Cook, *Halley*, 53.

34 See Willmoth, *Sir Jonas Moore*, 151–2.

35 cit. Ibid., 179–80.

36 Ibid., 169.

37 cit. Iliffe, 'Privacy, property and priority', 29.

38 cit. Willmoth, *Sir Jonas Moore*, 175.

39 Iliffe, 'Privacy, property and priority', 45.

40 Newton to Josiah Burchett, 26 August 1725, cit. Andrewes, 'Even Newton could be wrong', 190.

41 cit. Andrewes, 'Even Newton could be wrong', 191.

42 Cook, *Halley*, 166–7.

43 Willmoth, *Sir Jonas Moore*, 169.

44 See Espinasse, *Robert Hooke*, 61–3.

45 Willmoth, *Sir Jonas Moore*, 192.

46 Landes, *Revolution in Time*, 136.

47 Willmoth, *Sir Jonas Moore*, 193.

48 Francis Baily (1835); cit. W. J. Ashworth, 'John Flamsteed, property and intellectual labour', 201.

49 cit. Ashworth, 'John Flamsteed, property and intellectual labour', 213–4.

50 cit. Ibid., 212–3.

51 Cook, *Halley*, 392.

52 cit. Ashworth, 'John Flamsteed, property and intellectual labour', 216.

53 cit. van Helden, 'Longitude and the satellites of Jupiter', 97–8.

54 Van Helden, *The Quest for Longitude*, 99.

55 cit. Ashworth, 'John Flamsteed, property and intellectual labour', 207.

56 Iliffe, 'Privacy, property and priority', 52.

57 Landes, *Revolution in Time*, 135–8.

58 Ibid., 116.

CHAPTER 5: BREAKING NEW GROUND

1 Cook, *Halley*, 53.
2 Ibid., 86.
3 See above, chapter 4.
4 Flamsteed, *Corr.* I, 451, 595–7.
5 Cook, *Halley*, 66.
6 Gingerich, 'How astronomers finally captured Mercury', 39.
7 Cook, *Halley*, 74.
8 Ibid., 116.
9 Ibid., 114.
10 Cook, *Halley*, 117–25; Débarbat, 'Newton, Halley, and the Paris Observatory', 33–4
11 Cook, *Halley*, 116.
12 Leopold, 'The longitude timekeepers of Christiaan Huygens', 106–7.
13 Cook, *Halley*, 115.
14 Simon Schaffer, 'Newton's comets and the transformation of astrology', 225.
15 Brown, *The Story of Maps*, 245–6.
16 On the finances and politics of the Académie, see Stroup, *A Company of Scientists*.
17 Thrower, 'Longitude in the context of cartography', 55–6; *Maps and Civilization*, 110.
18 Thrower, *Maps and Civilization*, 110–13.
19 Brown, *The Story of Maps*, 254–5.
20 Van Helden, 'Longitude and the satellites of Jupiter'; Débarbat, S. and Wilson, C., 'The Galilean satellites of Jupiter from Galileo to Cassini, Rømer and Bradley', in Taton and Wilson (eds), *Planetary Astronomy from the Renaissance to the Rise of Astrophysics*, 144–57.
21 Stroup, *A Company of Scientists*, 53.
22 Guy de Tachard, *A Relation of the Voyage to Siam, Performed by Six Jesuits, Sent by the French King to the Indies and China in the Year 1685, with their Astrological Observations, and their Remarks of Natural Philosophy, Geography, Hydrography, and History* (London, 1688), I.
23 Ibid., 50.
24 Ibid., 52–3.
25 Ibid., 58.
26 Cook, *Halley*, 349, 503.
27 On Huygens's later longitude clocks see Leopold, 'The longitude time-

keepers of Christiaan Huygens', 108–11.

28 cit. Iliffe, 'Un pôle de controverses'.

29 Cook, *Halley*, 172.

30 Turnbull, 'Cartography and science in early modern Europe', 18–19.

31 K. Zandvliet, *Mapping for Money*; Koeman, *Joan Blaeu and his Grand Atlas*, 23–5.

32 Koeman, *John Blaeu*, 9–10; John Goss, 'Blaeu's The Grand Atlas of the 17th Century World' (Royal Geographic Society, 1990).

33 Tachard, *A Relation of the Voyage to Siam*, 6.

34 Koeman, *John Blaeu*, 51.

35 Taylor, 'Robert Hooke and the cartographical projects', 529.

36 Taylor, '"The English Atlas" of Moses Pitt'.

37 Hooke, *Diary*, 441, 458.

38 MacGregor, *Sir Hans Sloane*, 270.

39 Willmoth, *Sir Jonas Moore*, 130–1; see also E. M. G. Routh, *Tangier: England's Lost Atlantic Outpost* (London, 1912).

40 Bennett, *The Mathematical Science of Christopher Wren*.

41 For details of the fen drainage project, see Willmoth, *Sir Jonas Moore*, 88–120.

42 cit. Ibid., 74.

43 Ibid., 221.

44 Ibid., 195–207; Iliffe, 'Mathematical Characters'.

45 cit. Willmoth, *Sir Jonas Moore*, 109.

46 Ibid., 120.

47 Ibid., 134.

48 MacGregor (ed.), *Sir Hans Sloane*, 250.

49 Reproductions in Routh, *Tangier*.

50 Willmoth, *Sir Jonas Moore*, 184.

51 cit. Ibid., 135.

52 Ibid., 136.

53 Willmoth, *Sir Jonas Moore*, 133. See also E. Chappell (ed.), *The Tangier Papers of Samuel Pepys*, Navy Records Society 73 (1935).

54 Willmoth, *Sir Jonas Moore*, 136.

55 Ibid., 136.

56 Frank, *Harvey and the Oxford physiologists*, passim.

57 cit. Shapin, *The Social History of Truth*, 260.

58 Ibid., 261.

59 Cook, *Halley*, 238.

60 MacGregor (ed.), *Sir Hans Sloane*, 12.

61 Sir Hans Sloane, *A Voyage to the Islands Madera, Barbados, Nieves, S. Christophers and*

Jamaica, with the Natural History of the four-footed beasts, fishes, birds, insects, reptiles &c,
2 vols (London, 1707), lxxx–lxxxi.

62 Cook, *Halley*, 236–7.
63 Ibid., 238.
64 Ibid., 239.
65 Ibid., 240.
66 Ibid., 240–42.

CHAPTER 6: STRANGE SPECIMENS

 1 A. Pavord, *The Tulip* (London: Bloomsbury, 1999), 171.
 2 Tachard, *Voyage to Siam*, 51.
 3 cit. Heniger, *Hendrik Adriaan Van Reede Tot Drakenstein*, 72.
 4 Ibid., 72.
 5 MacGregor (ed.), *Sir Hans Sloane*, 142–3. On Hermann see Gunn and
    Codd, *Botanical Exploration of Southern Africa*, 26–7.
 6 cit. Karsten, *The Old Company's Garden at the Cape*, 57–8.
 7 Israel, *The Dutch Republic*, 908.
 8 cit. Heniger, *Hendrik Adriaan Van Reede Tot Drakenstein*, 37.
 9 cit. Ibid., 42.
10 cit. Ibid., 54.
11 Israel, *The Dutch Republic*, 909.
12 Wettengl (ed.), Maria Sybilla Merian, 250–53, 259–60; N. Z. Davis,
    Women on the Margins, 178–9.
13 Gunn and Codd, *Botanical Exploration of Southern Africa*, 250.
14 Tachard, *Voyage to Siam*, 63.
15 Gunn and Codd, *Botanical Exploration of Southern Africa*, 34.
16 De Beer, *Sir Hans Sloane*, 32–3.
17 Gunn and Codd, *Botanical Exploration of Southern Africa*, 28.
18 Wijnands, *The Botany of the Commelins*, 3–5.
19 See above chapter 3.
20 Blunt and Stearn, *The Art of Botanical Illustration*, 113.
21 Figures in Stroup, *A Company of Scientists*, appendices.
22 Ibid., 194.
23 Ibid., 39, 95.
24 Heniger, *Hendrik Adriaan Van Reede Tot Drakenstein*, 44.
25 Evelyn, *Diary*, 135.
26 G. S. Rousseau, 'Pineapples, pregnancy, pica, and *Peregrine Pickle*', 189.
27 Ibid., 193.

28   Bacon, 'Of Gardens'.

29   On the Tradescants see M. Allan, *The Tradescants*.

30   MacGregor, in Impey and MacGregor, *The Origins of Museums*, 149–50.

31   Hunter, 'Elias Ashmole', in *Science and the Shape of Orthodoxy*, 37.

32   On Ashmole see Hunter, 'Elias Ashmole', in *Science and the Shape of Orthodoxy*, and Josten, *Elias Ashmole*.

33   MacGregor (ed.), *Sir Hans Sloane*, 106.

34   Hooke, *Diary*, 288.

35   Mea Allan, *The Tradescants*, 215.

36   Hunter, *Elias Ashmole 1617–1692*, 57.

37   Hunter, 'Elias Ashmole', in *Science and the Shape of Orthodoxy*, 43.

38   MacGregor, in Impey and MacGregor, *The Origins of Museums*, 152.

39   MacGregor (ed.), *Sir Hans Sloane*, 12.

40   Ibid., 12.

41   Ibid., 93.

42   Sir Hans Sloane, *A Voyage to the Islands Madera, Barbados, Nieves, S. Christophers and Jamaica* 1, 6.

43   MacGregor (ed.), *Sir Hans Sloane*, 114.

44   Sir Hans Sloane, *A Voyage to the Islands Madera, Barbados, Nieves, S. Christophers and Jamaica* 2, 346.

45   MacGregor (ed.), *Sir Hans Sloane*, 13.

46   Sir Hans Sloane, *A Voyage to the Islands Madera, Barbados, Nieves, S. Christophers and Jamaica* 1, xx.

47   See MacGregor (ed.), *Sir Hans Sloane*, 278–90; Carrubba, *Engelbert Kaempfer*.

48   See MacGregor (ed.), *Sir Hans Sloane*, 142 for a page from Hermann's herbarium.

49   MacGregor (ed.), *Sir Hans Sloane*, 22–3.

50   Ibid., 24.

51   Ibid., 24.

52   On Sloane's will see ibid., 45–68.

53   On organising scientific papers see M. Hunter (ed.), *Archives of the Scientific Revolution*.

54   Principe, *The Aspiring Adept*, 11–26.

Chapter 7: Examining the Evidence

1   See Carrubba, *Engelbert Kaempfer*, 120–38.

2   See W. Michel, *Hermann Buschof – Erste Abhandlung ber die Moxibustion in Europa* (Heidelberg: [imprint], 1993).

3 Cook, 'Medicine and natural history', 50–1. On gout see Porter and Rousseau, *Gout: The Patrician Malady*.

4 Birch, cit. MacGregor, *Sir Hans Sloane*, 11.

5 On bezoars see MacGregor (ed.), *Sir Hans Sloane*, 71–3.

6 Shapin, *Social History of Truth*, 347–9.

7 cit. MacGregor (ed.), *Sir Hans Sloane*, 27.

8 Ibid., 130.

9 Stroup, *A Company of Scientists*, 212–13.

10 Ibid., 202–3.

11 Printed in K. Wettengl, *Maria Sybilla Merian*, 265.

12 Ibid., 269.

13 Jarcho, *Quinine's Predecessor*, 53.

14 See ibid., 44–58.

15 Gideon Harvey, cit. Jarcho, *Quinine's Predecessor*, 55.

16 MacGregor (ed.), *Sir Hans Sloane*, 15.

17 Sloane, *Natural History* I, cxxxiv.

18 cit. Porter and Rousseau, *Gout: The Patrician Malady*, 41.

19 Ibid., 44.

20 MacGregor (ed.), *Sir Hans Sloane*, 39.

21 Ibid., 244.

22 Foust, *Rhubarb*, 6–7.

23 Ibid., 46–55.

24 Ibid., 37.

25 Ibid., 30–41.

26 Ibid., 35–6.

27 Hooke, *Diary*, 11.

28 Ibid., 13.

29 Ibid., 14.

30 Ibid., 49–50.

31 cit. Debus *French Paracelsians*, 25.

32 Ibid., 21–2.

33 Hooke, *Diary*, 8.

34 Ibid., 172.

35 Beier, 'Experience and experiment: Robert Hooke, illness and medicine', 241–2.

36 Hooke, *Diary*, 99.

37 Jardine and Stewart, *Hostage to Fortune*, 507–8.

38 Foust, *Rhubarb*, 43.

39 Hooke, *Diary*, 67.

40 Ibid.,105.
41 *Micrographia*, fol. gir.
42 E. F. MacPike (ed.), *Correspondence and Papers of Edmond Halley* (London: Taylor and Francis, 1937), 217.
43 Hooke, *Diary*, 85.
44 *Micrographia*, 81–2.
45 Ibid., 144–5.
46 Porter and Rousseau, *Gout: The Patrician Malady*, chapter 5.
47 Debus, *French Paracelsians*, 89.
48 Cook, *Trials of an Ordinary Doctor*, 76–105.
49 Raven, *John Ray*, 228.
50 Ibid., 294–5.
51 Ibid., 304–5.
52 On Ray and Wilkins's *Real Character* see Raven, *John Ray*, 181–6; Shapiro, *John Wilkins*.
53 Raven, *John Ray*, 210–11.
54 Ibid., 211.
55 Ibid., 226–7.
56 Ibid., 301.
57 See J. L. Reveal, *Gentle Conquest: The Botanical Discovery of North America With Illustrations from the Library of Congress* (Library of Congress Classics) (Fulcrum Publishing, 1992); MacGregor (ed.), *Sir Hans Sloane*, 141.
58 See also J. L. Reveal website.
59 See chapter 2 above.
60 Raven, *John Ray*, 349–50.
61 Ibid., 365.
62 See chapter 5 above.
63 See chapter 1 above.
64 cit. Cook, *Halley*, 161.

Chapter 8: Committed to Paper

1 Ford, *Single Lens*, 59.
2 Hunter and Schaffer (eds), *Robert Hooke: New Studies*, 133.
3 Hooke, *Micrographia*, 133.
4 Ford, *Single Lens*, 33–59.
5 Birch, *History of the Royal Society* 3, 352.
6 Eamon, *Science and the Secrets of Nature*; 'From the secrets of nature to public knowledge'.

7 See above, chapter 4.

8 Hall and Hall, *Correspondence of Henry Oldenburg* 12, 14.

9 Ibid., 12–13.

10 Gunther 8, 150–1.

11 BL copy in the King's Library has Sloane's catalogue number (E6) on its fly-leaf. It may therefore be Hooke's copy, since Sloane bought other Hooke volumes at the 1703 auction of Hooke's library.

12 Gunther 8, 151.

13 Hooke, *Lectures De Potentia Restitutiva, or of Spring, Explaining the Power of Springing Bodies* (London: John Martyn Printer to the Royal Society, 1678), reprinted in Gunter 8, 331–88; 333.

14 Oldenburg, *Corr.*, 136–9.

15 J. V. Golinski, 'Chemistry in the Scientific Revolution', 379.

16 Newton, *Corr.* 3, 193.

17 Ibid., 215.

18 Ibid., 218. Cf. Newton to ——, undated; *Corr.* 7, 393.

19 Westfall, *Never at Rest*, 491–3.

20 The following account is closely based on Principe, *Aspiring Adept*, 155–180.

21 Ibid., 157.

22 Ibid., 157.

23 Newton, *Corr.* 3, 45.

24 Principe, *Aspiring Adept*, 162.

25 Ibid., 115.

26 Ibid., 116.

27 On the Pierre incident see ibid., 115–34.

28 Principe, *Aspiring Adept*, 176–9.

29 Leibniz to Huygens, 11 October 1693, cit. Leopold, 'Longitude Timekeeper of Christiaan Huygens', 102.

30 Maor, *e: The Story of a Number*, 117.

31 See above, chapter 2.

32 Ibid.

33 Birch *History of the Royal Society* 2, 984.

34 Shapin, *Social History of Truth*, 403.

35 See above, chapter 2.

36 See Shapin, *Social History of Truth*, 355–83.

37 Schiebinger, 'Maria Winkelmann', 311.

38 Ibid., 314.

39 Ibid., 317.

40 Cook, *Halley*, 101.

41 Ibid., 102.

42 Cit, Brusati, *Artifice and Illusion*, 109.

43 John Speed's *History of Great Britaine* (1650) and *Prospect of the most famous parts of the world* (1631), and Wagenaer's *Mariner's Mirrour* (1588); the 'book of Cards' was a collection of sea-charts.

44 Pepys, *Diary* 6, 290–2.

45 Dugdale's losses are described in his letters of 15 October and 28 May 1667 (*Life, diary and corr.*, ed. W. Hamper, 1827, 364).

46 See above, chapter 7.

47 Pepys, *Diary* 6, 309–10.

48 Birch, *Boyle Works* 6, 229.

49 Ibid., 230.

50 Birch, *History of the Royal Society* 3, 330–1.

51 Evelyn, *Diary*, 252.

52 Hooke, *Diary*, 270.

53 Ibid., 270.

54 Ibid., 277.

55 Ibid., 361–2, and 374 onwards.

56 Oldenburg, *Correspondence* 1, Introduction.

57 Ibid., Foreword, xvii–xviii.

58 Evelyn, *Diary*, 183.

59 See above, chapter 4.

60 Oldenburg, *Correspondence* 12.

EPILOGUE

1 Walter Sullivan, the *New York Times*, February 15, 1968.

2 Stent (ed.), *The Double Helix*, 233.

3 Ibid., 191.

4 Ibid., 193.

5 Schrödinger, *What is Life?*

6 Stent (ed.), *The Double Helix*, 31.

7 Ibid., 86.

8 Stent in ibid., xvi.

9 Ibid., 32–4.

10 Ibid., 98.

11 Ibid., 98.

12 On the anthropology of science, see Latour, *Laboratory Life*.

FURTHER READING

INTRODUCTION

Bronowski, J., *Science and Human Values* (London: Hutchinson, 1961)

Dawkins, R., *Unweaving the Rainbow* (London: Allen Lane The Penguin Press, 1998)

Jardine, L., *Worldly Goods: A New History of the Renaissance* (London: Macmillan, 1996)

Latour, B., *Science in Action: How to Follow Scientists and Engineers through Society* (London: Open University Press, 1987)

Rose, S., *Lifelines: Biology, Freedom, Determinism* (London: Allen Lane The Penguin Press, 1997)

Wolpert, L., *The Unnatural Nature of Science* (London: Faber and Faber, 1992)

CHAPTER I: SIGNS OF THE TIMES

Bechler, Z., *Newton's Physics and the Conceptual Structure of the Scientific Revolution* (Dordrecht: Kluwer Academic Publishers, 1991)

Bennett, J. A., "'On the power of penetrating into space": the telescopes of William Herschel,' *Journal for the History of Astronomy* 7 (1976), 75–108

Biagoli, M., *Galileo Coutier: The Practice of Science in the Culture of Absolutism* (Chicago: University of Chicago Press, 1993)

Curry, P. (ed.), *Astrology, Science and Society: Historical Essays* (Woodbridge: Boydell Press, 1987)

Feingold, M., 'Astronomy and strife: John Flamsteed and the Royal Society', in Willmoth (ed.), *Flamsteed's Stars*, 31–48

Field, J. V., *Kepler's Geometrical Cosmology* (Chicago: University of Chicago Press, 1988)

Genuth, S. S., *Comets, Popular Culture and the Birth of Modern Cosmology* (Princeton: Princeton University Press, 1997)

Hall, A. R., and Westfall, R. S., 'Did Hooke concede to Newton?' *Isis* 58 (1967), 403–5

Koestler, A., *The Sleepwalkers: A History of Man's Changing Vision of the Universe* (London: Hutchinson, 1959)

Rousseau, G. S., 'Wicked Whiston and the English wits', in *Enlightenment Borders*, 325–41

Sabra, A. I., *Theories of Light from Descartes to Newton* (2nd edn, Cambridge: Cambridge University Press, 1981)

Sagan, C., and Druyan, A., *Comet* (New York: Random House, 1985)

Schaffer, S., 'Newton's comets and the transformation of astrology', in Curry (ed.), *Astrology, Science and Society*, 219–43

Schaffer, S., 'Halley, Delisle, and the making of the comet', in Thrower (ed.), *Standing on the Shoulders of Giants*, 254–98

Westfall, R. S., *Never at Rest: A Biography of Isaac Newton* (Cambridge: Cambridge University Press, 1980)

Whitfield, P., *The Mapping of the Heavens* (London: British Library, 1995)

CHAPTER 2: SCALE MODELS

Bennett, J. A., 'Robert Hooke as mechanic and natural philosopher', *Notes and Records of the Royal Society* 35 (1980), 33–48

Bennett, J. A., *The Mathematical Science of Christopher Wren* (Cambridge: Cambridge University Press, 1982)

Drake, E. T., *Restless Genius: Robert Hooke and his Earthly Thoughts* (Oxford: Oxford University Press, 1996)

'Espinasse, M., *Robert Hooke* (London: Heinemann, 1956)

Field, J. V., and James, F. A. J. L. (eds), *Renaissance and Revolution: Humanists, Scholars, Craftsmen and Natural Philosophy in Early Modern Europe* (Cambridge: Cambridge University Press, 1993)

Ford, B. J., *Single Lens: The Story of the Simple Microscope* (New York: Harper and Row, 1985)

Hall, A. R., *Hooke's Micrographia, 1665–1965* (London, 1966)

Hooke, R., *Robert Hooke's Micrographia* (Lincolnwood, Illinois: Science Heritage Ltd., 1987)

Hunter, M., and Schaffer, S. (eds), *Robert Hooke: New Studies* (Woodbridge: Boydell Press, 1989)

Iliffe, R., 'Material Doubts: Hooke, artisan culture and the exchange of information in 1670s London', *British Journal for the History of Science* 28 (1995), 285–318

Ruestow, E. G., *The Microscope in the Dutch Republic: The Shaping of Discovery* (Cambridge: Cambridge University Press, 1996)

Sargent, R.-M., *The Diffident Naturalist: Robert Boyle and the Philosophy of Experiment* (Chicago: University of Chicago Press, 1995)

Schierbeek, A., *Measuring the Invisible World: The Life and Works of Antoni van Leeuwenhoek FRS* (London: Abelard-Schuman, 1959)

Shapin, S., and Schaffer, S., *Leviathan and the Air-Pump: Hobbes, Boyle, and the Experimental Life* (Princeton: Princeton University Press, 1985)

Wilson, C., *The Invisible World: Early Modern Philosophy and the Invention of the Microscope* (Princeton: Princeton University Press, 1995)

CHAPTER 3: SUBTLE ANATOMY

Alpers, S., *The Art of Describing: Dutch Art in the Seventeenth Century* (London: John Murray, 1983)

Brusati, C., *Artifice and Illusion: The Art and Writing of Samuel Van Hoogstraten* (Chicago: University of Chicago Press, 1995)

Cunningham, A., *The Anatomical Renaissance: The Resurrection of the Anatomical Projects of the Ancients* (Aldershot, Hants: Scolar Press, 1997)

Debus, A. G., *The French Paracelsians: The Chemical Challenge to Medical and Scientific Tradition in Early Modern France* (Cambridge: Cambridge University Press, 1991)

Fournier, M., *The Fabric of Life: Microscopy and the Seventeenth Century* (Baltimore: Johns Hopkins University Press, 1996)

Frank, R. G. Jr., *Harvey and the Oxford Physiologists: A Study of Scientific Ideas* (Berkeley: University of California Press, 1980)

Gaukroger, S., *Descartes: An Intellectual Biography* (Oxford: Clarendon Press, 1995)

Guyton, A. C. (ed.), and Willis, R. (transl.), *The Works of William Harvey* (Philadelphia: University of Pennsylvania Press, 1989) (original edition, 1965)

Hall, M. B., *Robert Boyle and Seventeenth-Century Chemistry* (Cambridge: Cambridge University Press, 1958)

Israel, J., *The Dutch Republic: Its Rise, Greatness, and Fall 1477–1806* (Oxford: Clarendon Press, 1995)

Kaplan, B. B., *'Divulging of Useful Truths in Physick': The Medical Agenda of Robert Boyle* (Baltimore: Johns Hopkins University Press, 1993)

Lawrence, C., and Shapin, S. (eds), *Science Incarnate: Historical Embodiments of Natural Knowledge* (Chicago: University of Chicago Press, 1998)

Lenoble, R., *Mersenne, ou la naissance du méchanisme* (Paris: Vrin, 1943)

Lindeboom, G. A., *Het Cabinet van Jan Swammerdam (1637–1680)* (Amsterdam, 1980)

Miller, J., *On Reflection* (London: National Gallery Publications Ltd., 1998)

Pagel, W., 'Harvey and the purpose of the circulation', *Isis* 42 (1951), 22–38

Schaffer, S., 'Regeneration: the body of natural philosophers in Restoration England', in Lawrence and Shapin (eds), *Science Incarnate*, 83–120

Shapiro, B. J., *John Wilkins 1614–1672: An Intellectual Biography* (Berkeley and Los Angeles: University of California Press, 1969)

Verbeek, T., *Descartes and the Dutch: Early Reactions to Cartesian Philosophy, 1637–1650* (Carbondale and Edwardsville: Southern Illinois University Press, 1985)

Wear, A., 'William Harvey and the "way of the anatomists"', *History of Science* 21 (1983), 223–49

CHAPTER 4: RUNNING LIKE CLOCKWORK

Andrewes, W. J. H., 'Even Newton could be wrong: the story of Harrison's first three sea clocks', in Andrewes (ed.), *The Quest for Longitude*, 190–234

Andrewes, W. J. H., 'Finding local time at sea, and the instruments involved', in Andrewes (ed.), *The Quest for Longitude*, 394–404

Andrewes, W. J. H. (ed.), *The Quest for Longitude* (Cambridge, Mass.: Collection of Historical Scientific Instruments, Harvard University, 1996)

Ashworth, W. J., '"Labour harder than *thrashing*": John Flamsteed, property and intellectual labour in nineteenth-century England', in Willmoth (ed.), *Flamsteed's Stars*, 199–216

Bedini, S. A., *The Pulse of Time: Galileo Galilei, the determination of longitude, and the pendulum clock* (Florence: Leo S. Olschki, 1991)

Berlin, I., *Many Thousands Gone: The First Two Centuries of Slavery* (New York: Bellnap Press, 1998)

Bos, H. M. J. *et al* (eds), *Studies on Christiaan Huygens: invited papers from* the *Symposium on the Life and Work of Christiaan Huygens, Amsterdam, 22–25 August 1979* (Lisse, The Netherlands: Swets & Zeitlinger, 1980)

Burke, J. G. (ed.), *The Uses of Science in the Age of Newton* (Berkeley: University of California Press, 1983)

Crosby, A. W., *The Measure of Reality: Quantification and Western Society, 1250–1600* (Cambridge: Cambridge University Press, 1997)

Defossez, L., *Les savants du XVIIe siècle et la mesure du temps* (Lausanne: Edition du Journal Suisse d'Horlogerie et de Bijouterie, 1946)

Dohrn-van Rossum, G., *The History of the Hour: Clocks and Modern Temporal Orders* (Dunlap, T., transl.) (Chicago: University of Chicago Press, 1996)

Hall, A. R., 'Horology and criticism: Robert Hooke', *Studia Copernicana* 16 (1978), 261–81

Hall, A. R., 'Mechanics and the Royal Society 1668–1670', *British Journal for the History of Science* 3 (1966), 24–38

Hall, A. R., 'Robert Hooke and Horology', *Notes and Records of the Royal Society* 8 (1951), 167–77

Hall, A. R., and Hall, M. B., 'Why blame Oldenburg?' *Isis* 53 (1962), 482–91

Howse, D., 'The lunar-distance method of measuring longitude', in W. J. H. Andrewes (ed.), *The Quest for Longitude*, 150–62

Howse, D., *Greenwich Observatory: Vol. 3. Buildings and Instruments* (London: Taylor and Francis, 1975)

Howse, D., *Greenwich Time and the Discovery of the Longitude* (Oxford: Oxford University Press, 1980)

Howse, D. (ed.), *Francis Place and the Early History of The Greenwich Observatory* (New York, 1975)

Iliffe, R., '"In the warehouse": privacy, property and priority in the early Royal Society', *History of Science* 30 (1992), 29–68

Kokott, W., 'Astronomische Längenbestimmung in der frühen Neuzeit', *Sudhoffs Arch.* 79 (1995), 165–72

Laudan, L., 'The clock metaphor and probabilism: the impact of

Descartes on English methodological thought, 1650–65,' *Annals of Science* 22 (1966), 73–104

Leopold, J. H., 'Christiaan Huygens and his instrument makers,' in Bos *et al* (eds), *Studies on Christiaan Huygens*, 221–33

Leopold, J. H., 'The longitude timekeepers of Christiaan Huygens', in Andrewes (ed.), *The Quest for Longitude*, 102–14

Mahoney, M. S., 'Christiaan Huygens: The measurement of time and longitude at sea', in Bos *et al* (eds.), *Studies on Christiaan Huygens*, 234–70

Price, D. J., 'The manufacture of scientific instruments from c. 1500 to c. 1700,' in C. Singer *et al* (eds), *A History of Technology*, (London: Oxford University Press, 1957) vol. 3, 620–47

Pumfrey, S., 'Ideas above his station: a social study of Hooke's curatorship of experiments', *History of Science* 29 (1991), 1–44

Schaffer, S., 'Astronomers mark time: discipline and the personal equation', *Science in Context* 2 (1988), 115–45

Sobel, Dava, *Longitude: The True Story of a Lone Genius Who Solved the Greatest Scientific Problem of His Time* (London: Fourth Estate, 1995)

Stimson, A., 'The longitude problem: the navigator's story', in W. J. H. Andrewes (ed.), *The Quest for Longitude*, 72–84

van Helden, A., 'Longitude and the satellites of Jupiter', in Andrewes (ed.), *The Quest for Longitude*, 86–100

von Mackensen, L., von Bertele, H., and Leopold, J. H., *Die erste Sternwarte Europas mit ihren Instrumenten und Uhren: 400 Jahre Jost Bürgi in Kassel* (Munich: Callwey, 1979)

Willmoth, F., *Flamsteed's Stars: New Perspectives on the Life and Work of the First Astronomer Royal (1646–1719)* (Woodbridge: Boydell Press 1997)

Yeo, R., 'Genius, method, and morality: images of Newton in Britain, 1760–1860', *Science in Context* 2 (1988), 257–84

CHAPTER 5: BREAKING NEW GROUND

Brotton, J., *Trading Territories: Mapping the Early Modern World* (London: Reaktion Books, 1997)

Brown, L. A., *The Story of Maps* (Boston: Little, Brown, 1949)

Brown, L. A., *Jean Domenique Cassini and his World Map* (Ann Arbor: University of Michigan Press, 1941)

Cook, A., *Edmond Halley: Charting the Heavens and the Seas* (Oxford: Clarendon Press, 1998)

Débarbat, S., 'Newton, Halley and the Paris Observatory', in Thrower, *Standing on the Shoulders of Giants*, 27–52

Débarbat, S., and Wilson, C., 'The Galilean satellites of Jupiter from Galileo to Cassini, Rømer and Bradley', in Taton and Wilson (eds), *Planetary Astronomy from the Renaissance to the Rise of Astrophysics*, 144–57

Dickinson, H. W., *Sir Samuel Morland, Diplomat and Inventor, 1625–1695* (Cambridge: Cambridge University Press, 1970)

Fara, P., *Sympathetic Attractions: Magnetic Practices, Beliefs, and Symbolism in Eighteenth-Century England* (Princeton: Princeton University Press, 1996)

Gingerich, O., 'How astronomers finally captured Mercury', in *The Great Copernicus Chase*, 36–42

Gingerich, O., *The Great Copernicus Chase and Other Adventures in Astronomical History* (Cambridge: Cambridge University Press, 1992)

Goldgar, A., 'De part et d'autre de la Manche', in J.-P. Isicovics (ed.), *Cahiers de Science et Vie* (June 1998), 90–6

Goss, J., 'Blaeu's The Grand Atlas of the 17th Century World', *Royal Geographical Society*, 1990

Greenberg, J. L., *The Problem of the Earth's Shape from Newton to Clairaut: The Rise of Mathematical Science in Eighteenth-Century Paris and the Fall of 'Normal' Science* (Cambridge: Cambridge University Press, 1995)

Howse, D., *Nevil Maskelyne: The Seaman's Astronomer* (Cambridge: Cambridge University Press, 1989)

Iliffe, R., 'Auto-evaluation et la mecanique céleste selon Newton' in J.-P. Isicovics (ed.), *Cahiers d'Histoire de la Vie et des Sciences* (March 1993), 40–68

Iliffe, R., 'Ce que Newton connut sans sortir de chez lui', *Histoire et Mesure* 4 (1994), 1–41

Iliffe, R., 'Maupertius, precision measurement and the expedition to Lapland in the 1730s', *History of Science* 31 (1993), 335–75

Iliffe, R., 'Un pôle de controverses', in J.-P. Isicovics (ed.) *Cahiers de Science et Vie* (June 1998), 56–63

Koeman, C., *Joan Blaeu and his Grand Atlas* (Amsterdam: Theatrum Orbis Terrarum Ltd., 1970)

Konvitz, J., *Cartography in France, 1660–1848: Science, Engineering, and Statecraft* (Chicago: University of Chicago Press, 1987)

Lux, D. S., *Patronage and Royal Science in Seventeenth-Century France: The Académie de Physique in Caen* (Ithaca: Cornell University Press, 1989)

Nunis, D. B., *The 1769 Transit of Venus: The Baja California Observations* (Los Angeles: Natural History Museum of Los Angeles County, 1982)

Olmsted, J. W., 'The scientific expedition of Jean Richer to Cayenne (1672–1673)', *Isis* 34 (1942)

Reinhartz, D., *The cartographer and the literati: Herman Moll and his intellectual circle* (Lewiston, NY: E. Mellen Press, 1997)

Rostenberg, L., 'Moses Pitt, Robert Hooke and the *English Atlas*', *The Map Collector* 12 (1980), 2–8

Schwartz, S. J., and Ehrenberg, R. E., *The Mapping of America* (New York: Harry N. Abrams, Inc., 1980)

Shapin, S., *A Social History of Truth: Civility and Science in Seventeenth-Century England* (Chicago: University of Chicago Press, 1994)

Smith, J. M., *The Culture of Merit: Nobility, Royal Service, and the Making of Absolute Monarchy in France, 1600–1789* (Ann Arbor: University of Michigan Press, 1996)

Taton, R., and Wilson, C. (eds), *Planetary Astronomy from the Renaissance to the Rise of Astrophysics* (Cambridge: Cambridge University Press, 1989)

Taylor, E. G. R., '"The English Atlas" of Moses Pitt, 1680–83', *The Geographical Journal* 95 (1940), 292–9

Taylor, E. G. R., 'Robert Hooke and the cartographical projects of the late seventeenth century (1666–1696)', *The Geographical Journal* 90 (1937), 529–40

Thrower, N. J. W., 'Longitude in the context of cartography', in Andrewes (ed.), *The Quest for Longitude*, 50–62

Thrower, N. J. W., *Maps and Civilization: Cartography in Culture and Society* (Chicago: University of Chicago Press, 1972; revised edn 1996)

Thrower, N. J. W., *Original Survey and Land Subdivision* (1966)

Thrower, N. J. W. (ed.), *Standing on the Shoulders of Giants: A Longer View of Newton and Halley* (Berkeley, University of California Press, 1990)

Thrower, N. J. W. (ed.), *The Three Voyages of Edmond Halley in the Paramore 1698–1701* (London: The Hakluyt Society, 1981)

Turnbull, D., 'Cartography and science in early modern Europe: mapping the construction of knowledge spaces', *Imago Mundi* 48 (1996), 5–24

Turner, A. J., 'In the wake of the Act, but mainly before', in Andrewes (ed.), *The Quest for Longitude*, 116–32

Waters, D. W., 'Captain Edmond Halley, FRS, Royal Navy, and the practice of Navigation', in Thrower (ed.), *On the Shoulders of Giants*, 171–202

Willmoth, F., 'Mathematical sciences and military technology: the Ordnance Office in the reign of Charles II', in Field and James, *Renaissance and Revolution*, 117–31

Willmoth, F., *Sir Jonas Moore: Practical Mathematics and Restoration Science* (Woodbridge: Boydell Press, 1993)

Wood, P. H., 'La Salle: Discovery of a lost explorer', *American Historical Review* 89 (1984), 294–323

Zandvliet, K., *Mapping for Money: Maps, Plans and Topographical Paintings and their Role in Dutch Overseas Expansion during the Sixteenth and Seventeenth Centuries* (Amsterdam: Batavian Lion International, 1998)

CHAPTER 6: STRANGE SPECIMENS

Allan, M., *The Tradescants: Their Plants, Gardens and Museum 1570–1662* (London: Michael Joseph, 1964)

Bann, S., *Under the Sign: John Bargrave as Collector, Traveler, and Witness* (Ann Arbor: University of Michigan Press, 1994)

Beekman, E. M. (ed. and transl.), *The Poison Tree: Selected Writings of Rumphius on the Natural History of the Indies* (Amherst: University of Massachusetts Press, 1981)

Beer, G. R. de, *Sir Hans Sloane and the British Museum* (London: Oxford University Press for the British Museum, 1953)

Bermingham, A., and Brewer, J., *Consumption of Culture; 1600–1800; Image, Object, Text* (London: Routledge, 1997)

Blunt, W., and Stearn, W. T., *The Art of Botanical Illustration*, 2nd edn (Woodbridge: Antique Collector's Club, 1994)

Brewer, J., *The Pleasures of the Imagination: English Culture in the Eighteenth Century* (London: HarperCollins, 1997)

Dandy, J. E., *The Sloane Herbarium* (London, 1958)

Dixon Hunt, J., '*Curiosities* to adorn *Cabinets* and *Gardens*', in Impey and MacGregor, *The Origins of Museums*, 193–203

Findlen, P., *Possessing Nature: Museums, Collecting, and Scientific Culture in Early Modern Italy* (Berkeley: University of California Press, 1994)

Goffman, D., *Britons in the Ottoman Empire 1642–1660* (Seattle: University of Washington Press, 1998)

Gunn, M. E., and Codd, L. E., *Botanical Exploration of Southern Africa* (Netherlands: AA Balkema, 1981)

Hamel, J., *England and Russia: comprising, The voyages of John Tradescant the elder, Sir Hugh Willoughby, Richard Chancellor, Nelson, and others to the*

White Sea (transl. John Studdy Leigh) (New York: Da Capo Press, 1968: Facsimile reprint first published 1854)

Heniger, J., *Hendrik Adriaan Van Reede Tot Drakensteing and Hortus Malabaricus: A Contribution to the History of Dutch Colonial Botany* (Rotterdam: Balkema, 1986)

Hunter, M. (ed.), *Archives of the Scientific Revolution: The Formation and Exchange of Ideas in Seventeenth-Century Europe* (Woodbridge: Boydell and Brewer, 1998)

Hunter, M., 'The Cabinet institutionalized: The Royal Society's "Repository" and its background', in Impey and MacGregor, *The Origins of Museums*, 159–68

Hunter, M., *John Aubrey and the Realm of Learning* (New York: Science History Publications, 1975)

Hunter, M., *Elias Ashmole 1617–1692. The Founder of the Ashmolean Museum and his World: A Tercentenary Exhibition* (Oxford: Ashmolean Museum, 1983)

Hunter, M., 'Mapping the mind of Robert Boyle', in Hunter (ed.), *Archives of the Scientific Revolution*, 121–36

Hunter, M., *Science and the shape of orthodoxy: intellectual change in late seventeenth-century Britain* (Woodbridge: Boydell Press, 1995)

Impey, O., and MacGregor, A. (eds), *The Origins of Museums: The Cabinet of Curiosities in Sixteenth- and Seventeenth-Century Europe* (Oxford: Clarendon Press, 1985)

Jardine, N., Secord, J. A., and Spary, E. C. (eds), *Cultures of Natural History* (Cambridge: Cambridge University Press, 1996)

Josten, C. H., 'Elias Ashmole FRS (1617–1692)', *Notes and Records of the Royal Society of London* 15 (1960), 221–30

Josten, C. H. (ed.), *Elias Ashmole. His Autobiographical and Historical Notes, his Correspondence, and other Contemporary Sources relating to his Life and Work*, 5 vols (Oxford: Clarendon Press, 1966)

Karsten M., *The Old Company's Garden at the Cape* (Cape Town, 1951)

LeFanu, W., *Nehemiah Grew MD FRS: A Study and Bibliography of his Writings* (Winchester: St. Paul's Bibliographies, 1990)

Leith-Ross, P (ed.), *The John Tradescants: Gardeners to the Rose and Lily Queen* (London: Peter Owen, 1984)

Ludwig, H., *Nürnberger naturgeschichtliche Malerei im 17. und 18 Jahrhundert* (Marburg: Basilisken-Presse, 1997)

MacGregor, A., 'The cabinet of curiosities in seventeenth-century Britain', in Impey and MacGregor, *The Origins of Museums*, 147–58

MacGregor, A. (ed.), *Sir Hans Sloane: Collector, Scientist, Antiquary, Founding Father of the British Museum* (London: British Museum Press, 1994)

MacGregor, A. (ed.), *Tradescant's Rarities: Essays on the Foundation of the Ashmolean Museum, 1683, with a Catalogue of the Surviving Early Collections* (Oxford: Clarendon Press, 1983)

Ovenell, R. F., *The Ashmolean Museum 1683–1894* (Oxford: Clarendon Press, 1986)

Pollock, L., *With Faith and Physic: The Life of a Tudor Gentlewoman, Lady Grace Mildmay 1552–1620* (London: Collins and Brown, 1993)

Prest, J. M., *The Garden of Eden: The Botanic Garden and the Re-Creation of Paradise* (New Haven: Yale University Press, 1982; p/b edition 1988)

Raven, C. E., *English Naturalists from Neckham to Ray: A Study of the Making of the Modern World* (Cambridge: Cambridge University Press, 1947)

Reveal, J. L., *Gentle Conquest: The Botanical Discovery of North America With Illustrations from the Library of Congress* (Library of Congress Classics) (Fulcrum Publishing, 1992)

Rousseau, G. S., 'Pineapples, pregnancy, pica, and *Peregrine Pickle*', in *Enlightenment Borders*, 176–99

Rousseau, G. S., *Enlightenment Borders: Pre- and Post-Modern Discourses, Medical, Scientific* (Manchester: Manchester University Press, 1991)

Stroup, A., 'Académie royale ou parisienne?', in J.-P. Isicovics (ed.) *Cahiers de Science et Vie* (June 1998), 22–30

Stroup, A., *A Company of Scientists: Botany, Patronage, and Community at the Seventeenth-Century Parisian Royal Academy of Sciences* (Berkeley:

University of California Press, 1990)

Thomas, K., *Man and the Natural World: Changing Attitudes in England 1500–1800* (London: Penguin Books, 1991)

Welch, Martin, *The Tradescants and the Foundation of the Ashmolean Museum* (Oxford: Ashmolean Museum Publications, 1978)

Wettengl, K. (ed.), *Maria Sybilla Merian 1647–1717* (London: Thames and Hudson, 1998)

Wijnands, D. O., *The Botany of the Commelins* (Rotterdam: Balkema, 1983)

Wright, D., *Elias Ashmole: founder of the Ashmolean Museum, Oxford, Archaeologist, Astrologer, Historian, Rosicrucian, and Freemason* (London: The Freemason, c. 1910)

CHAPTER 7: EXAMINING THE EVIDENCE

Beier, L. M., 'Experience and experiment: Robert Hooke, illness and medicine', in Hunter and Schaffer (eds), *Robert Hooke: New Studies*, 235–52

Beier, L. M., *Sufferers and Healers: The Experience of Illness in Seventeenth-century England* (London, 1987)

Brockliss, L., and Colin Jones, *The Medical World of Early Modern France* (Oxford: Clarendon Press, 1997)

Carrubba, R. W. (ed. and transl.), *Engelbert Kaempfer: Exotic Pleasures* (Carbondale: Southern Illinois University Press, 1996)

Christie, J. R. R., and Golinski, J. V., 'The spreading of the word: new directions in the historiography of chemistry 1600–1800', *History of Science* 20 (1982), 235–66

Cook, H. J., 'The new philosophy and medicine in seventeenth-century England', in Lindberg and Westman (eds), *Reappraisal of the Scientific Revolution*, 397–436

Cook, H. J., *Trials of an Ordinary Doctor: Joannes Groenevelt in Seventeenth-century London* (Baltimore: Johns Hopkins University Press, 1994)

Cook, H. J., 'The cutting edge of a revolution? Medicine and natural history near the shores of the North Sea', in Field and James (eds), *Renaissance and Revolution*, 45–61

Cook, H. J., *The Decline of the Old Medical Regime in Stuart London* (Ithaca: Cornell University Press, 1986)

Davis, N. Z., *Women on the Margins: Three Seventeenth-century Lives* (Cambridge, Mass.: Harvard University Press, 1995)

Dear, P., 'A mechanical microcosm: bodily passions, good manners, and Cartesian mechanism', in Lawrence and Shapin (eds), *Science Incarnate*, 51–82

Dear, P., *Mersenne and the Learning of the Schools* (Ithaca: Cornell University Press, 1988)

Debus, A. G., 'The Paracelsian aerial niter', *Isis* 55 (1964), 43–61

Debus, A. G., *The Chemical Philosophy: Paracelsian Science and Medicine in the Sixteenth and Seventeenth Centuries*, 2 vols (New York: Science History Publications, 1977)

Dewhurst, K., 'Some letters of Dr Charles Goodall to Locke, Sloane and Millington', *Journal of the History of Medicine* 17 (1962), 487–508

Douglas, Mary, 'Self-evidence', in *Implicit Meanings: Essays in Anthropology* (London: Routledge and Kegan Paul, 1975), 276–318

Eamon, W., '"With the rules of life and an enema": Leonardo Fioravanti's medical primitivism', in Field and James (eds), *Renaissance and Revolution*, 29–44

Eamon, W., 'From the secrets of nature to public knowledge', in Lindberg and Westman (eds), *Reappraisals of the Scientific Revolution*, 333–65

Eamon, W., *Science and the Secrets of Nature: Books of Secrets in Medieval and Early Modern Culture* (Princeton: Princeton University Press, 1994)

Foust, C. M., *Rhubarb: The Wondrous Drug* (Princeton: Princeton University Press, 1992)

French, R. K., and Wear, A. (eds), *Medical Revolution in the Seventeenth Century* (Cambridge: Cambridge University Press, 1989)

Guitton, G., *Les Jesuites à Lyon sous Louis XIV et Louis XV* (Lyon, 1953)

Gwei-Djen, L., *Celestial Lancets: History and Rationale of Acupuncture and Moxa* (Cambridge: Cambridge University Press, 1980)

Jarcho, S., *Quinine's Predecessor: Franceso Torti and the Early History of Cinchona* (Baltimore: Johns Hopkins University Press, 1993)

Jardine, L., and Stewart, A., *Hostage to Fortune: The Troubled Life of Francis Bacon* (London: Gollancz, 1998)

Kaufmann, T. da C., *The Mastery of Nature: Aspects of Art, Science and Humanism in the Renaissance* (Princeton: Princeton University Press, 1993)

McKie, D., 'Fire and *flamma vitalis*: Boyle, Hooke and Mayow,' in *Science, Medicine and History*, ed. E. A. Underwood, 2 vols., vol. I, 469–88 (London: Oxford University Press, 1953)

Michel, W., *Hermann Buschof – Erste Abhandlung ber die Moxibustion in Europa* (Heidelberg, 1993)

Needham, J., *The Grand Titration: Science and Society in East and West* (London: George Allen & Unwin, 1969)

Pagel, W., *Joan Baptista Van Helmont: Reformer of Science and Medicine* (Cambridge: Cambridge University Press, 1982)

Pagel, W., *Paracelsus: an introduction to philosophical medicine in the era of the Renaissance* (London, 1982)

Pagel, W., *The Smiling Spleen: Paracelsianism in Storm and Stress* (Basel: Karger, 1984)

Porter, R., *Health for Sale: Quackery in England 1650–1850* (Manchester: Manchester University Press, 1989)

Porter, R., *The Greatest Benefit to Mankind: A Medical History of Humanity from Antiquity to the Present* (London: HarperCollins, 1997)

Porter, R., and Rousseau, G. S., *Gout: The Patrician Malady* (New Haven: Yale University Press, 1998)

Sawday, J., *The Body Emblazoned: Dissection and the Human Body in Renaissance Culture* (London: Routledge, 1995)

Schaffer, S., 'Self evidence', *Critical Inquiry* 18 (1992), 327–62

Schivelbusch, W., transl. Jacobson, D., *Tastes of Paradise: A Social History of Spices, Stimulants, and Intoxicants* (New York: Columbia University Press, 1992)

Smith, P. H., *The Business of Alchemy: Science and Culture in the Holy Roman Empire* (Princeton: Princeton University Press, 1994)

Steele, A. R., *Flowers for the King: the expedition of Ruiz and Pavon and the Flora of Peru* (Durham, N. C.: Duke University Press, 1964)

Strasser, G. F., 'Closed and open languages: Samuel Hartlib's involvement with cryptology and universal languages', in Greengrass, Leslie and Raylor, *Samuel Hartlib and Universal Reformation*, 151–61

CHAPTER 8: COMMITTED TO PAPER

Boas, M., *Robert Boyle and Seventeenth-Century Chemistry* (Cambridge: Cambridge University Press, 1958)

Chandler, J., Davidson, A. I., and Harootunian, H. (eds), *Questions of Evidence: Proof, Practice, and Persuasion Across the Disciplines* (Chicago: University of Chicago Press, 1994)

Clericuzio, A., 'Carneades and the chemists: a study of *The Sceptical Chymist* and its impact on seventeenth-century chemistry, in Hunter (ed.), *Robert Boyle Reconsidered*, 79–89

Collins, H. M., *Changing Order: Replication and Induction in Scientific Practice* (London: Sage, 1985)

Darnton, R., *The Business of Enlightenment: A Publishing History of the 'Encyclopédie', 1775–1800* (Cambridge, Mass.: Harvard University Press, 1978)

Dear, P. (ed.), *The Scientific Enterprise in Early Modern Europe: Readings from Isis* (Chicago: University of Chicago Press, 1997)

Eisenstein, E. L., *The Printing Press as an Agent of Change: Communications and Cultural Transformations in Early-Modern Europe* (Cambridge: Cambridge University Press, 1980)

Evans, R. J. W., *Rudolf II and his World: A Study in Intellectual History 1576–1612* (Oxford: Clarendon Press, 1973)

Feingold, M., *The Mathematicians' Apprenticeship* (Cambridge: Cambridge University Press, 1984)

Gaukroger, S., *Explanatory Structures: Concepts of Explanation in Early Physics and Philosophy* (Atlantic Highlands, NJ: Humanities Press, 1978)

Golinski, J. V., 'Chemistry in the scientific revolution: problems of language and communication', in Lindberg and Westman (eds), *Reappraisals of the Scientific Revolution*, 367–96

Golinski, J., *Making Natural Knowledge: Constructivism and the History of Science* (Cambridge: Cambridge University Press, 1998)

Grafton, A., *Commerce with the Classics: Ancient Books and Renaissance Readers* (Ann Arbor: University of Michigan Press, 1997)

Grant, E., *Much Ado About Nothing: Theories of Space and Vacuum from the Middle Ages to the Scientific Revolution* (Cambridge: Cambridge University Press, 1981)

Hacking, I., *The Emergence of Probability: A Philosophical Study of Early Ideas about Probability, Induction and Statistical Inference* (Cambridge: Cambridge University Press, 1975)

Howarth, D., *Lord Arundel and his Circle* (New Haven: Yale University Press, 1985)

Hunter, M. (ed.), *Robert Boyle Reconsidered* (Cambridge: Cambridge University Press, 1994)

Hunter, M., and Wood, P. B., 'Towards Solomon's House: Rival strategies for reforming the early Royal Society', *History of Science* 24 (1986), 49–108

Hunter, M., 'A "College" for the Royal Society: the abortive plan of 1667–1668', *Notes and Records of the Royal Society* 38 (1984), 159–86

Hunter, M., 'Promoting the new science: Henry Oldenburg and the early Royal Society', *History of Science* 25 (1988), 165–81

Iliffe, R., and Willmoth, F., 'Domestic Science: Margaret Flamsteed and Caroline Herschel as assistant-astronomers', in Hunter, L., and Hutton, S. (eds), *Women, Science and Medicine 1500–1700* (Stroud, Glos: Sutton Publishing, 1997), 235–65

Iliffe, R., 'Isaac Newton: Lucatello Professor of Mathematics', in C. Lawrence and S. Shapin (eds), *Science Incarnate*, 121–55

Iliffe, R., 'Mathematical Characters: Flamsteed and Christ's Hospital Royal Mathematical School', in F. Willmoth (ed.) *Flamsteed's Stars: New Perspectives* (Bristol, 1997), 115–44

Jacob, M. C., *The Newtonians and the English Revolution, 1689–1720* (Ithaca: Cornell University Press, 1976)

Jardine, L., *Francis Bacon: Discovery and the Art of Discourse* (Cambridge: Cambridge University Press, 1974)

Long, P. O., 'Invention, authorship, "intellectual property", and the origin of patents: notes toward a conceptual history', *Technology and Culture* 32 (1991), 846–84

Mahoney, M. S., *The Mathematical Career of Pierre de Fermat 1601–1665* (Princeton: Princeton University Press, 1973; second edition 1994)

Maor, E., *e: The Story of a Number* (Princeton: Princeton University Press, 1994)

Mendelson, S. H., *The Mental World of Stuart Women* (Brighton: Harvester Press, 1987)

Moran, B. T. (ed.), *Patronage and Institutions: Science, Technology, and Medicine at the European Court 1500–1750* (Woodbridge: Boydell Press, 1991)

Peck, L. L., 'Uncovering the Arundel Library at the Royal Society', *Proceedings of the Royal Society* (1998)

Popkin, R. H., *The History of Scepticism from Erasmus to Spinoza* (Berkeley: University of California Press, 1979)

Principe, L. M., 'Apparatus and reproducibility in alchemy', in *Instruments and Experimentation in the History of Chemistry* (Chicago: University of Chicago Press, 1999)

Principe, L. M., 'Boyle's alchemical pursuits', in Hunter (ed.), *Robert Boyle Reconsidered*, 91–105

Principe, L. M., 'Robert Boyle's alchemical secrecy: codes, ciphers, and concealments', *Ambix* 39 (1992), 63–74

Principe, L. M., 'The alchemies of Robert Boyle and Isaac Newton: alternate approaches and divergent deployments', in M. J. Osler (ed.), *Canonical Imperatives: Rethinking the Scientific Revolution* (Cambridge: Cambridge University Press, 1999)

Principe, L. M., *The Aspiring Adept: Robert Boyle and his Alchemical Quest* (Princeton: Princeton University Press, 1998)

Rosen, E. (transl. and commentary), *Nicholas Copernicus: Minor Works* (Baltimore: Johns Hopkins University Press, 1985)

Rousseau, G. S., 'Science books and their readership in the high enlightenment', in *Enlightenment Borders*, 264–324

Schiebinger, L., 'Maria Winkelmann at the Berlin Academy: a turning point for women in science', in Dear (ed.), *The Scientific Enterprise*, 305–31

Schmitt, C. B., 'Experience and experiment: a comparison of Zabarella's view with Galileo's in *De motu*,' *Studies in the Renaissance* 16 (1969), 80–137

Schmitt, C. B., 'Experimental evidence for and against a void: the sixteenth-century arguments,' *Isis* 58 (1967), 352–66

Shapin, S., *The Scientific Revolution* (Chicago: University of Chicago Press, 1996)

Shapiro, B. J., *Probability and Certainty in Seventeenth-century England: A Study of the Relationships between Natural Science, Religion, History, Law, and Literature* (Princeton: Princeton University Press, 1983)

Szydlo, Z., 'The influence of the central nitre theory of Michael Sendivogius on the chemical philosophy of the seventeenth century', *Ambix* 43 (1996), 80–96

Teeter Dobbs, B. J., *The Janus Face of Genius: The Role of Alchemy in Newton's Thought* (Cambridge: Cambridge University Press, 1991)

Thomas, K., 'Numeracy in early modern England', *Transactions of the Royal Historical Society* 37 (1987) 103–32

van Leeuwen, H. G., *The Problem of Certainty in English Thought 1630–1690* (The Hague: Nijhoff, 1963) von Leyden, W., *Seventeenth-century Metaphysics: An Examination of Some Main Concepts and Theories* (London: Duckworth, 1968)

Wallis, R. (ed.), *On the Margins of Science: The Social Construction of Rejected Knowledge*, Sociological Review Monograph No. 27 (Keele: Keele University Press, 1979)

Webster, C., *From Paracelsus to Newton: Magic and the Making of Modern Science* (Cambridge: Cambridge University Press, 1982)

Webster, C., *The Great Instauration: Science, Medicine and Reform 1626–1660* (London: Duckworth, 1975)

Yoder, J. G., *Unrolling Time: Christiaan Huygens and the Mathematization of Nature* (Cambridge: Cambridge University Press, 1984)

EPILOGUE

Bragg, M. with Gardiner, R., *On Giants' Shoulders: Great Scientists and their Discoveries from Archimedes to DNA* (London: Hodder & Stoughton, 1998)

Crick, F., *What Mad Pursuit: A Personal View of Scientific Discovery* (Harmondsworth: Penguin Books, 1990)

Jones, S., *The Language of the Genes* (London: HarperCollins, 1991)

Latour, B., *Laboratory Life: The Social Construction of Scientific Facts* (Beverly Hills, Calif.: Sage, 1979)

Sayre, A., *Rosalind Franklin and DNA* (New York: W. W. Norton & Co., 1978)

Schrödinger, E., *What is Life? The Physical Aspect of the Living Cell* (first published 1944) (Cambridge: Cambridge University Press, 1967)

Stent, G. S. (ed.), *The Double Helix: A Personal Account of the Discovery of the Structure of DNA, by James D. Watson* (Norton Critical Edition) (New York: W. W. Norton & Co., 1980)

GENERAL

Beer, E. S. de, *The Diary of John Evelyn* 6 vols (Oxford: Clarendon Press, 1955)

Birch, T. (ed.), *The History of the Royal Society of London* 4 vols (London, 1756–57)

Birch, T. (ed.), *The Works of the Honourable Robert Boyle* 6 vols (London, 1772)

Burtt, E. A., *The Metaphysical Foundations of Modern Physical Science* (revised. edn Garden City, NY: Anchor, 1954; orig. publ. 1924)

Cohen, H. F., *The Scientific Revolution: A Historiographical Inquiry* (Chicago: University of Chicago Press, 1994)

Cohen, I. B., and Westfall, R. S. (eds), *Newton: Texts, Backgrounds, Commentaries* (New York: W. W. Norton & Co., 1995)

Forbes, E. G., Murdin, D., and Willmoth, F. (eds), *The Correspondence of John Flamsteed* (vol. I) (Bristol and Philadelphia: The Institute of Physics Publishing, 1995)

Goodman, D., and Russell, C. A., *The Rise of Scientific Europe 1500–1800* (London: Hodder and Stoughton, 1991)

Gunther, R. T., *Early Science in Oxford* (Oxford: 'Printed for the Author', 1930)

Hall, A. R., *From Galileo to Newton 1630–1720* (London: Collins, 1963)

Hall, A. R., *The Revolution in Science 1500–1750* (London: Longman, revised edn 1983)

Hall, A. R., and Hall, M. B. (ed. and transl.), *The Correspondence of Henry Oldenburg* 13 vols: vols I–9 (Madison: University of Wisconsin Press, 1965–73); vols I0–II (London: Mansell, 1975–6); vols I2–I3 (London: Taylor & Francis, 1986)

Huff, T. E., *The Rise of Early Modern Science: Islam, China, and the West* (Cambridge: Cambridge University Press, 1993)

Hunter, M., *Establishing the New Science: The Experience of the Early Royal Society* (Woodbridge: Boydell Press, 1989)

Hunter, M., *The Royal Society and Its Fellows 1660–1700: The Morphology of an Early Scientific Institution* (Chalfont St. Giles: British Society for the History of Science, 1982)

Hunter, M. (ed.), *Archives of the Scientific Revolution: The Formation and*

Exchange of Ideas in Seventeenth-century Europe (Woodbridge: Boydell Press, 1998)

Huygens, C., *Oeuvres complètes de Christiaan Huygens* 22 vols (The Hague: Nijhoff, 1888–1950)

Kuhn, T. S., *The Essential Tension: Selected Studies in Scientific Tradition and Change* (Chicago: University of Chicago Press, 1977)

Kuhn, T. S., *The Structure of Scientific Revolutions* (Chicago: University of Chicago Press, revised edn 1970)

Lindberg, D. C., and Westman, R. S. (eds), *Reappraisal of the Scientific Revolution* (Cambridge: Cambridge University Press, 1990)

MacPike, E. F. (ed.), *Correspondence and Papers of Edmond Halley* (Oxford: Clarendon Press, 1932)

Munby, A. N. L. (ed.), *Sale Catalogues of Libraries of Eminent Persons* (London: Mansell Publishing Ltd., 1975)

Rigaud, S. J. (ed.), *Correspondence of Scientific Men of the Seventeenth Century* 2 vols (Oxford: The Clarendon Press, 1891)

Robinson, H., and Adams, W. (eds), *The Diary of Robert Hooke MA MD FRS 1672–1680* (London: Taylor and Francis, 1935)

Turnbull, H. W., Scott, J. F., Hall, A. R., and Tilling, L. (eds), *The Correspondence of Isaac Newton* 7 vols (Cambridge: Cambridge University Press, 1959–77)

Westfall, R. S., *The Construction of Modern Science: Mechanisms and Mechanics* (Cambridge: Cambridge University Press, 1977)

INDEX

Page numbers in *italic* refer to the illustrations

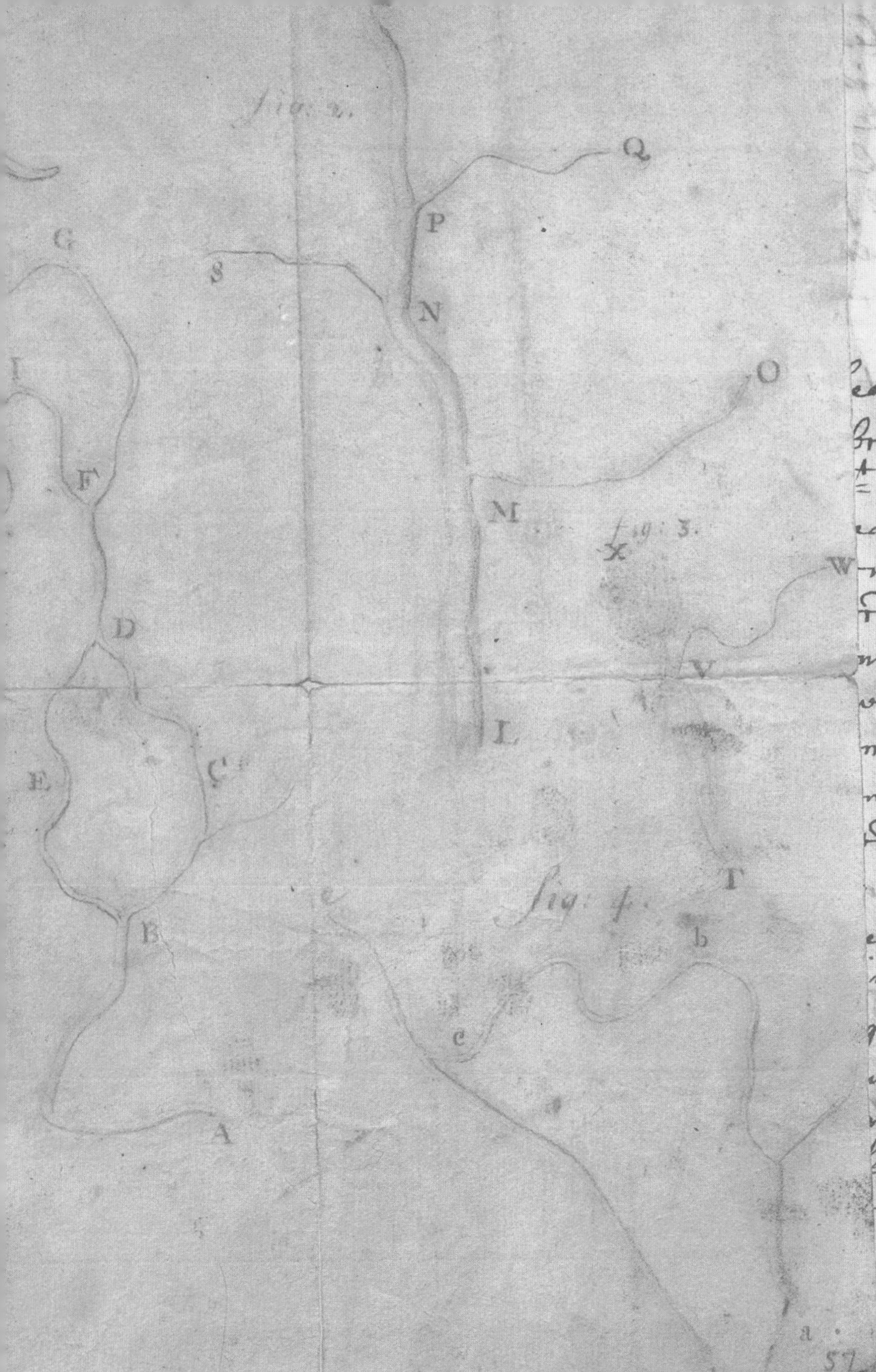

fig. 2.

G

I

F

D

E   C

B

A

Q

P

N

O

M   fig. 3.
X

W

V

L

T
b

fig. 4.

c

a
57
d

Hans Sloane
Secr: vande Roij: Societ

Hoog geleerde Heer